The Voice
of Genius

The Voice of Genius

Conversations with Nobel Scientists and Other Luminaries

Denis Brian

Perseus Publishing
Cambridge, Massachusetts

Cataloging-in-Publication Data is available from
the Library of Congress

ISBN 0-7382-0447-1

Perseus Publishing is a member of the Perseus Books Group.

Find us on the World Wide Web at
http://www.perseuspublishing.com

Perseus Publishing books are available at special discounts
for bulk purchases in the U.S. by corporations, institutions,
and other organizations. For more information, please contact
the Special Markets Department at HarperCollins Publishers,
10 East 53rd Street, New York, NY 10022, or call 1-212-207-7528.

First paperback printing, January 2001

1 2 3 4 5 6 7 8 9 10—03 02 01

To Martine, Danielle, and Alex, with love.

Preface

Why?

Almost everything is still a mystery: the why, when, and how of the universe and its inhabitants. The vagaries of human behavior and subatomic particles seem equally inscrutable. Is our world an incredible accident or has it a purpose? What is there left to discover? Are our brains—made from the same atoms as stars—simply sophisticated computers or something more?

God not being available I looked for scientists on the cutting edge, the brightest and most daring thinkers of our time, many of them Nobel prize winners. I wanted to know their latest findings in trying to unravel the secrets of the universe from its creation to its destiny. I hoped to gain insight into the workings of the human brain and psyche, to discuss what still intrigues and mystifies them, to get a sense of their lives and personalities, and to learn of their close encounters with Einstein; Bohr; Heisenberg [Did Nazi Germany's greatest scientist try to build an atom bomb for Hitler or to sabotage the effort?]; his American counterpart, Oppenheimer; Freud; and Jung.

Even though I was obviously a layman in all their disciplines—chemistry, physics, astronomy, biology, anthropology, psychiatry, psychology, psychoanalysis, and neuroscience—all except one agreed to talk with me. He had other engagements but sent his latest publication.

My conversations with Linus Pauling form the first chapter in this book. Just before he died, a former colleague charged him with stealing his ideas and sued Pauling and his Institute. I checked the outcome, and you will find this in the update at the end of the book.

More recently newspapers reported drastically shorter estimates for the age of the universe given by a team of astronomers using the repaired Hubble Space Telescope. I checked this with veteran astronomer Allan Sandage, who learned astronomy from Hubble himself and discovered quasars in his quest to find out how the world is put together. Nicknamed "Super-Hubble," Sandage has been called "the only person on Earth fully versed in observational cosmology."* His response is also in the last chapter.

Here, then, are some of the greatest minds of our century sharing their ideas and discussing their fascinating discoveries and experiences.

---

* *The Red Limit*, Timothy Ferris, Morrow, 1977, p. 109.

Acknowledgments

Many thanks to: Doron Weber, Rockefeller University; Ann Hughes, Chandler Medical Center; Bob Ford and Julia Heiney, AT&T Bell Laboratories; Nina Murray; Olivia Anonas, University of California, San Diego; Dorothy Munro, Linus Pauling Institute; Patricia Orr, Caltech; Victor Swoboda, Montreal Neurological Institute; Cornell University.

Special thanks to my wife, Martine who gave me invaluable help and many of the best questions.

And to my editor, Linda Greenspan Regan, for her advice and enthusiasm for the project.

Contents

Contents

Chapter One.   *Linus Pauling*

"I have often astounded myself"
—Linus Pauling

Asked during sworn testimony to name the greatest physical chemist, a University of Chicago professor replied, "Linus Pauling." When reproved for choosing someone who'd been called a fraud and a Communist stooge, the professor replied, "I was under oath. I had to tell the truth."[1]

Pauling didn't only excel in chemistry. He also majored in controversy, and was surely unmatched in his mastery of physics, molecular biology, immunology, genetic diseases, metallurgy and medicine (psychiatry included), not to mention nuclear weaponry and radioactive fallout. The *New Scientist* ranked him, with Einstein and Newton, among the top twenty of history's greatest scientists. He was unique among them in winning two unshared Nobel prizes—one for chemistry and one for peace. He deserved a third, say fans, for his medical breakthroughs. Had critics awarded him a prize, it might have been to the scientist who most enraged the establishment. Some even considered him a traitor. My aim was to find out who was closest to the truth: friends or enemies. I also wanted to hear how he made his discoveries, to invite him to speculate about unsolved cosmic and earthbound mysteries. To pick his brain, in fact, as well as probe his character and personality. I spoke with him at length in 1988. Over the years I corresponded with him and we had our most recent conversation on March 14, 1993, shortly after his 92nd birthday. Eighteen months later, on August 19, 1994, he died of cancer. He was 93.

Linus Pauling's redwood deck was perched precariously over the Pacific ocean. If he fell, or the deck collapsed, there would be no one within miles to help him, except for his caretaker who lived in a nearby house. At times Pauling couldn't even be reached by phone because the fog closing in from the ocean corroded the line and shorted the circuit. But he liked the isolation of his 160-acre estate, even though it once meant spending 24 hours trapped on a cliff with rocks and the raging sea far below.

At 92 he felt pretty good after a rocky year when he learned he had prostate cancer, in remission when we spoke. He woke most mornings alone in his Big Sur cabin to the sound of waves, made his own breakfast, then went to work with hand calculator and notebook, "trying to explain complicated phenomena on simpler scientific principles." With this in mind, Pauling was revising three books and investigating the cardiovascular system—among other pursuits.

Did he thrive on tension and conflict as one biographer, Anthony Serafini, [*Linus Pauling: The Man and His Science*] believed? He certainly provoked them by unusually successful forays from his specialty, chemistry, into physics, medicine, nutrition, psychiatry, even nuclear weaponry, annoying the hell out of rival practitioners—not to mention his controversial political activities.

Pauling's father, an Oregon druggist, died of peritonitis at 32. Linus, then nine, hardly knew him and turned for a role model to his maternal grandfather, Linus Darling, a family legend and the epitome of courage, enterprise and perseverance, whose own family broke up when he was eleven. Darling became "bound" to strangers who made him work for food and shelter. But when they told him he'd have to sleep in a barrel, he ran away. He was taken in by a more compassionate family, taught himself to read and write, and became in turn teacher, land surveyor, postmaster, and finally a lawyer and a justice of the peace. One of young Linus' aunts, a famous safecracker, though never with criminal intent, came at the call of those who had lost keys or forgotten combinations. A less practical uncle communed with the spirit of an Indian named Red Cloud, vainly hoping to get him to reveal the way to a lost gold mine.

Like his grandfather Linus was daring, enterprising, and persistent. And like his aunt became a sort of safe cracker, though on a cosmic scale—unlocking nature's secrets. He had little in common with the spiritualist uncle and grew to have less. To Pauling the spirit world is a myth, God included.

This skepticism was reinforced in his youth. He was looking at a drawing of a head of Christ, "when I saw a halo appear above it, a band of light. After investigation I discovered the after-image effects of the retina and satisfied myself that this was a general natural phenomenon."[2]

His schooldays foreshadowed his future. Insatiably curious, he was either glued to a book or scouring the countryside for his collection of insects and minerals. He also scared the neighbors by testing explosive chemicals on a nearby trolley track, the trolley acting as the trigger when coming into contact with the chemicals. At 13 he spoke some Greek and German and could count to a hundred in Chinese. He has continued to shock, annoy, or delight, ever since.

At Oregon State University Linus used scientific insight even while emptying his bladder. His roommate, Harold Granrud, could hardly forget how they were serenading girls. It was pitch dark, lots of trees around. Girls sleeping in porches upstairs whispering, "wondering who it was down there. Linus turned to me and said: 'Granny, I just confirmed a scientific fact here; the angle of incidence is greater than the angle of refraction, so I can piss against this tree without making any noise!' "[3]

In his senior year Pauling was asked to teach a chemistry course to 25 young women. He picked one at random to answer his first question. She was Ava Helen Miller, a pretty blonde with strong opinions, who turned out to be the brightest student in the class. He soon realized he was falling in love with her and, to avoid favoritism, gave her lower marks than justified. She was furious but sufficiently forgiving to marry him in 1923 when he was 22 and she 19.

At California Institute of Technology, Pauling was among the first to investigate the molecular structure of crystals. It was pioneering work, because, as he proudly notes, "all of our detailed knowledge of the structure of organic compounds, the dimensions and shapes, has come since that time when I began as a graduate student." He took a novel approach, by sending x-ray beams through the crystal. It worked like a charm, the atoms revealing their shape and position as a pattern on a photographic plate. By 1925, at 24, he had a doctorate with high honors and a three-month-old son, Linus Jr. (to be followed by two more sons and a daughter). A year later Pauling took off for Europe to study the strange new quantum mechanics. He worked with experts Arnold Sommerfeld, Erwin Schrodinger, and Niels Bohr. Unfortunately, says Pauling, "Bohr usually spoke in a mixture of English and other languages, and he was very hard to understand because he didn't pronounce the words easily."

But the trip paid off. Back home in December 1930, Pauling worked all through one night combining quantum mechanics with x-ray diffraction to determine what held atoms together. To his delight he found a simple mathematical method that solved difficult equations in minutes. "I kept getting more and more euphorious as time went by," he recalled, "and it didn't take me long to write a long paper about the nature of the chemical bond. That was a great experience," and a revolutionary approach to chemistry, eventually leading to new insight into the structure and behavior of all matter, not only of minerals and crystals but of proteins in the human body, such as the muscles, hair, and fingernails.

It was to win him the Nobel prize, and the admiration of his peers for a daring use of physics (quantum mechanics) to explain chemistry, which led to the new science of molecular biology. His immediate reward was promotion to professor at Caltech, and by 1937 he was chairman of the chemistry department. As an extension of his previous work he began to tackle the problem of how proteins—known to consist of strings of molecules—were joined together to produce fibrous structures such as the hair.

Until now he had taken little interest in politics. As a registered Republican he had first voted for Herbert Hoover, and later considered himself a Roosevelt democrat. His wife, however, was a radical, an enthusiastic member of the ACLU and the Women's International League for Peace and Freedom, who once described Norman Thomas, the Socialist candidate for the presidency, as "a bourgeois reactionary."[4] Pauling loved and admired his wife and was inclined to see things through her eyes.

This political move to the left was accelerated by a serious illness. In 1940 he was stricken with Bright's diseases (nephritis) and attended by a Dr. Addis who put him on a meatless, low-protein diet and vitamins, including a daily dose of 100 grams of vitamin C. Pauling believed the doctor saved his life and was impressed by the concern he had shown. In turn he sympathized when the doctor, an admitted Communist, was hounded by the State Department as a security risk. But there is little doubt it was Pauling's wife who turned him sharply and energetically to the left.

I asked if he might have been persuaded to move to the right had he married a political conservative, but instead of the expected "Of course not," Pauling said, "I don't answer hypothetical questions." If this sounds blunt, it wasn't. His tone was pleasant. He was simply stating a fact.

When Japanese-Americans were herded in internment camps after Pearl Harbor, Mrs. Pauling, with her husband's support, continued to employ their Japanese-American gardener. They were subjected to anonymous phone threats and letters and someone painted in red on their garage door, "Americans die, but we love Japs." But they kept the gardener on. And he escaped internment by joining the U.S. army.

During World War II, Robert Oppenheimer invited Pauling to run the Manhattan Project's chemistry section. But he was preoccupied with other war work: on explosives, rocket propulsion, immunology, and substitutes for human blood. After the war he focused on blood. He discovered that sickle-cell anemia, common among American blacks, is a "molecular disease" caused by an abnormality in the molecular structure of the red blood cells, preventing the free flow of oxygen. No one had previously suspected the existence of a molecular disease.

A virus and a dull detective story led to another of Pauling's major achievements soon afterwards. In 1948, bedridden with a heavy cold in Oxford, England, he sought solace in a whodunit. But on the second day he didn't give a damn whodunit. To pass the time he made what looked like doodles on a scrap of paper. After folding the paper, like a boy making a paper airplane, Pauling discovered one of the secrets of life, the alpha helix. This illustrates what he had puzzled over for years, how nature binds together chains of amino acids with helical (spiral) structures. Then, again by combining quantum mechanics with x-ray diffraction, he came up with a new theory of the nature of the chemical bond. It was the first big step toward unravelling the structure of DNA (the double helix).

But instead of following it up, he became a peace advocate. He agreed with Einstein that because rockets could now kill a million people and destroy cities, wars had become irrational and a new system needed to settle national disputes. Pauling's wife persuaded him to help prevent a catastrophe that would make all his work useless. So he joined a committee chaired by Einstein to campaign against nuclear weapons. Mrs. Pauling felt slightly guilty for diverting her husband from the scientific work he loved, but he soon came to share her passionate commitment. It was tough going. The American public was cool to the cause, especially after the Russians exploded an atom bomb in 1949. The committee also failed to prevent the test explosion of a hydrogen bomb in America in 1952 and in the Soviet Union a year later by which time mutual distrust had hardened into the Cold War.

Because of his antiwar activities the State Department suspected Pauling of being a Communist. In 1952 he was denied a passport and prevented from attending scientific conferences abroad. This may have jeopardized his chance of discovering the double helix: photographic clues to its structure were in England. Pauling demanded a hearing to establish his loyalty, and was exonerated. Despite this, it was still touch-and-go whether he'd get a passport for Sweden to collect his 1954 Nobel prize for chemistry. Fortunately the State Department relented.

In 1957, with the added prestige of the Nobel prize, Pauling told students at Washington University, St. Louis: "I believe that no human being should be sacrificed to the project of perfecting nuclear weapons that could kill hundreds of millions of human beings and that could devastate this beautiful world in which we live." The following year he launched a massive campaign, persuading 11,021 scientists from 49 nations to sign a petition urging the major powers to stop testing nuclear weapons as a prelude to general nuclear disarmament. "The Pauling Appeal" caused a sensation when presented to the United Nations. Opponents, notably Edward Teller, believed Pauling's plan was too dangerous because the Russians couldn't be trusted.

I asked Pauling why, despite both of them having access to the same political and technical information, their views were diametrically opposed.

He replied: "I see Edward Teller so rarely, but from newspaper accounts of interviews with him it looks as though we're not so far apart on nuclear power plants. He's worried about them, and suggested they all be buried underground. So far as nuclear weapons go we're still diametrically opposed. These are opinions about political decisions. Sometimes misrepresentations are made about the facts, but most of the differences of opinion are just a difference of opinion. I also had a confrontation with Edward Teller (on TV). I say we can trust the Russians. He says we can't. I say if we did trust them we wouldn't be in too great a danger, because we could always rearm again quickly enough to be safe. Also, it isn't necessary to trust the other country, because detection methods are so powerful now."

Science writer Isaac Asimov viewed Pauling not as a political radical, but as "a brave man who struggled for peace and against nuclear weapons even during the McCarthy era, when so many worthy men were frightened and browbeaten into silence . . . I'm sure Pauling's November 14, 1958, article in *Science* that discussed the dangers of carbon-14 played its part in the eventual agreement on the part of the three chief nuclear powers to suspend atmospheric testing."[5]

Pauling was pleased that the figures he gave in his 1953 book *No More War* estimating potential atom-bomb damage and radiation were still valid some 40 years later. "I made what I thought were the best estimates within a fivefold factor," he explained. "I said that possibly my numbers are five times too large or five times too small. The present figures are somewhat smaller than mine, perhaps a third less. But that doesn't change their impact significantly. The world would still be destroyed."

"In that field, as in others, there's a lot of misinformation. A government agency brought out a report about the effects of nuclear war. If you read it carefully the committee of scientists asked to make the report were told to ignore completely the effects of blast and fire and immediate fallout. So they responded in an honest way as to the effects of damage they were asked to assess. It's entirely unrealistic, because the major causes of death and other damage are ignored." And, Pauling concluded, that is lying by omission.

His peace campaign took him around the world in search of support. In Africa, in 1959, Albert Schweitzer signed the petition and broadcast pleas for others to join the cause.

When Pauling returned home, he disappeared. His wife reported him missing on January 31, 1960, and some news accounts suggested he might be dead. I asked him what happened.

Pauling: I was stuck on a cliff for 24 hours from ten in the morning until ten in the morning next day. I was what's called "ledged," about a hundred feet above the ocean and rocks below. I'd just scrambled across some loose scree and got on a ledge [while inspecting his land near Big Sur]. I couldn't go on and was afraid to go back. That's what climbers call "being ledged." I was irritated with myself. Climbers often go on a ledge and see it's a dangerous position and are afraid to go forwards or backwards. I was pretty sure I'd be rescued eventually. But I needed to stay awake all night so I wouldn't fall the hundred feet to the rocks below. I devised various ways. I tried to visualize the periodic table and state the names of the elements backwards, and I moved my arms and legs to keep warm. I used a map I had with me as a blanket to conserve body heat. And at night I tried to tell the time by using a stick to measure the movements of the constellations.

When I hadn't returned to our cabin by nightfall my wife went to the rangers for help. There weren't many people around that area but about 20 people were out looking for me and there may have been a helicopter. (The *Los Angeles Times* said a battery of forest rangers, sheriff's deputies, bloodhounds, and two H-189 Army helicopters from Fort Ord joined the search.) Radio reports early next morning announced I was missing and even speculated that I might be dead. About ten that morning someone spotted me on the ledge. The deputy sheriff then found a route that we could follow to safety. He came to just about where I was and directed me to clamber up the hill.

Someone sent you a Boy Scout whistle soon after, so you could whistle for help. Do you still have it?

Pauling: I don't know. I have one my son, Peter, gave me recently. I also received advice from an Englishwoman to get myself a couple of dogs to look after me. I didn't take her advice.

Pauling had hardly recovered from his cliffside ordeal when he was ordered to face an Internal Security Subcommmittee headed by Senator Dodd in 1960. Hoping to expose his peace work as Communist-inspired propaganda, a senator asked him to name all who helped collect the signatures of scientists opposed to nuclear weapons. He refused even though he feared he and his wife might be imprisoned for contempt of the Senate. But he wasn't charged with contempt, a tribute perhaps to his growing prestige.

He briefly returned to his scientific work to study the brain. In his 1961 paper, "A Molecular Theory of Anesthesia," he concluded that hundreds of substances can cause general anesthesia by turning water in the brain into submicroscopic ice-like crystals which shut off the electrical activity necessary for consciousness.

Early in 1962, "to keep my wife's respect," Pauling stepped up his efforts to prevent a potential nuclear holocaust. He urged an international ban on tests which contaminate the atmosphere and he feared would cause millions of children to be born with gross physical or mental defects. He took his appeal to the streets by first picketing the United Nations, then joined a woman's peace group outside the White House, carrying a sign: "Mr. Kennedy, Mr. Macmillan: We Have No *Right* to Test." With typical aplomb Pauling then joined other Nobelists at a White House reception where Kennedy congratulated him for continuing to express his opinions so forcibly. However, Mrs. Kennedy was less enthusiastic, asking him, "Why do you do that? Every time Caroline sees people outside with signs she says, 'What's daddy done *now*?'"[6]

The next year, alarm over reports of escalating radioactive fallout prompted President Kennedy to sign a test-ban treaty. That same day Pauling's Nobel peace prize was announced. It enraged his critics and *Life* magazine called it "an insult" to the U.S.

In 1962, the year Pauling had collected signatures for his peace petition, James Watson and Francis Crick got the Nobel prize for discovering the structure of the heredity molecule, DNA.

I asked; How close were you to finding the double helix?

Pauling: It's very hard to say. I had an example of the alpha helix, which I might well have discovered in 1937 when I worked on the problem. But, in fact, even then neither I nor anyone else discovered it for 11 years. (Pauling's discovery of the alpha helix, as mentioned, was the major first step toward unravelling the structure of DNA.) It was a pretty straightforward matter of discovery. It involved just one new idea that I hadn't used in 1937. Sir Lawrence Bragg, John Kendrew, and Max Perutz didn't discover it either, and they'd made a great effort to solve the problem of protein structure.

Some believe Watson and Crick cheated their colleague, Rosalind Franklin, out of the Nobel prize, by using her valuable discoveries and not sharing theirs with her. Though, as Pauling said, "It was her death (at 37) that cheated her." She died before the nominations were under consideration.

Pauling: I read the book about her, *Rosalind Franklin and DNA*,[7] and thought it persuasive that they had almost stolen material from her. The author even criticized me, but that was because of my ignorance of what was going on. I remained ignorant for a much longer time than I should have. The author says I should have reacted, I should have checked out the literature earlier. But I lost interest in the field after the double helix was discovered. I got interested in other things, and so I didn't go back and try to find out what the facts were. And perhaps memory failed me, too: I wrote an article and didn't mention Rosalind Franklin. I think the book about her is reliable and I don't think the statements have been refuted. It's too bad that she died before the Nobel committee made its assessment. She would surely have been awarded it along with Watson and Crick, but they rarely award them posthumously. If someone has been nominated before the first of February; say that the nomination is being considered but he dies before the prizes are awarded, he can be awarded the prize.

Were Watson and Crick unscrupulous as regards Rosalind Franklin?

Pauling: I don't want to summarize my feelings in a single adjective. Anything I said about them would have to be expressed in a greater number of words.

Does much of this happen in the scientific world?

Pauling: I don't think so. There probably are some scientists who are opportunists, who are eager to make their mark in the world and consider that more important than learning about nature. Crick isn't that sort of man at all. He has great interest in scientific problems and works away at them essentially independent of whether the solutions will benefit him. Crick is a good scientist. In *The Double Helix* you can see Watson is very much interested in fame and fortune. I think he's interested in science, too. He's very smart and his textbook on genetics is excellent.

When I asked Watson if from his youth he'd been aiming for the Nobel prize, he said, "Every boy thinks that if he hits an oilfield he's going to be wealthy, just as every young scientist feels that if he makes some great discovery he's going to get the Nobel prize."

Pauling: I don't know how true that is. I don't feel that every boy thinks about hitting oilfields and being wealthy.

Pauling's politics and principles brought him death threats. And William F. Buckley branded him "a megaphone of Soviet policy, and one of the nation's leading fellow-travelers." Buckley's *National Review* had targeted Pauling as someone who collaborated "with the enemy who has sworn to bury us . . . Year after year since time immemorial he has given his name, energy, voice and pen to one after another Soviet-serving enterprise."

Pauling sued for libel and asked for a million dollars in damages. Buckley defended anti-Pauling editorials as "true and published without malice."

At the trial Buckley's lawyer accused Pauling of supporting 93 "Soviet-serving and communist-aiding" organizations and individuals, such as The Communist Party for Young Communist League, the World Peace Conference, 11 convicted Communist leaders, and a conference of inquiry into the facts of the case against convicted atom bomb spies Ethel and Julius Rosenberg. Additional evidence indicating guilt by association was a 1961 photo of Pauling with Mrs. Khrushchev in Moscow, in which he seemed very cheerful. He was also quoted as saying if Kennedy resumed nuclear tests he would "go down to history as one of the most immoral men of all time and one of the greatest enemies of the human race," and that Truman was "irrational, ignorant or unscrupulous, or any combination of each."

In rebuttal, Pauling said the *National Review* had falsely branded him "a traitor, a collaborator with subversive foreign and alien elements . . . engaged in subversive Communist activities" and "a moral nihilist." Pauling claimed he had simply supported groups he thought would help "the cause of international understanding and world peace." Asked if he fully understood the Communist movement, atheist Pauling replied, "I was taught full understanding is possessed only by God."

Buckley also resorted to sophistry when asked what steps he took to ensure the charges against Pauling were true. Instead of producing tangible evidence, he said, "I filtered it through my memory, my understanding of contemporary history, my understanding of philosophy, my understanding of the decent relationships between people, my understanding of my religion, my understanding of the competence of my associates."

"What about your understanding of the facts?" asked Pauling's attorney.

"That's included in the list I gave you."

"But it wasn't mentioned."

"History is full of facts," said Buckley.

Not exactly Buckley's confident take on history, when he and I discussed the subject some years later.

Then I asked Buckley: Do you think it's a mistake to try to assess contemporary history or the men in control before most of the facts have been obtained?

Buckley: Lindbergh was widely condemned in this country for having accepted the Iron Cross from Hitler in 1937. It wasn't until 1954 or 1955 that it was revealed he was under a presidential request to do so, in effect for intelligence reasons. On the other hand, it's simply, as a human situation, impossible to wait . . . You don't know when history's going to come in with incremental datum and one has to make contingent judgments simply as a matter of necessity. For example, you have to know who to vote for.[8]

What Pauling didn't mention at the trial was that in 1962 he signed a petition asking an end to executions in the Soviet Union for "economic crimes."

After six weeks of testimony, Judge Samuel Silverman dismissed Pauling's complaint and presumably saved *National Review* from extinction. However, Silverman did "not hold that the charges against Dr. Pauling in the articles are true or justified . . . In all his actions Dr. Pauling acted well within his legal rights. And if his conscience required him to take the actions and pursue the course of conduct that he has pursued for the last 20 years, then he has acted in accordance with his moral duty. [He] has added the prestige of his reputation to aid the causes in which he believes. I merely hold that by so doing he also limited his legal remedies for any claimed libel of his reputation. And perhaps this can be deemed another sacrifice that he is making for the things he believes in."

Surprisingly, Pauling told me he agreed with the verdict.

Pauling: Justice Hugo Black favored very much that people should be protected against libel suits. The *New York Times* versus Sullivan suit settled part of the matter, saying a public servant can't sue for libel. Here was I, Linus Pauling. I wasn't a public servant, but I was a well-known public figure. And he felt that *New York Times* versus Sullivan should be extended to public figures. That was a good decision. I didn't like it. I didn't like having my libel suit fail. [And Pauling could have used the million bucks!] On the other hand, people who want to discuss important questions need to be free to mention what other well-known people have said and perhaps to criticize them. If I want to say I firmly disagree with President Clinton I don't want to have him filing a libel suit against me.

Are you pleased Clinton is president?

Pauling: Yes.

Having lost that libel suit do you feel restricted in what you say and do?

Pauling: No, I don't think so.

My attempts to get Pauling to clear the political air was like shadow boxing: good exercise but there wasn't much to show for it. When I asked if he had, in fact, supported communist causes, I might have been questioning Wittgenstein.

Pauling: I don't know what you mean by "support."

Didn't you sign your name or contribute to those causes?

Pauling: You'll have to be more specific.

Feeling uncomfortably like a latter day Joe McCarthy I tried again asking Pauling: Didn't you ever give your support or signature to organizations such as the Young Communist League?

Pauling: I'm not sure that I did.

Looking from another person's viewpoint, if you were told someone had apparently supported many causes with the word Communist in their titles, wouldn't you suspect they were Communist sympathizers?

Pauling: Here again, that's rather a vague and general question. I don't want to say yes or no to a question that doesn't have much meaning.

I took another tack: How did you respond when McCarthy called you a Communist stooge?

Pauling: I pointed out that I'd supported the right of Gerald K. Smith to speak freely. And that Smith was a far-right bigot.

How do you answer those who persist in the accusation?

Pauling: I usually quote Senator Hennings of Missouri, whom I overheard saying, "Pauling doesn't follow the Communist line, the Communists follow his line."

You certainly didn't follow the U.S. government line in 1984, did you?

Pauling: No, I was helping to escort $2.5 million-worth of supplies to Nicaragua aboard a Norwegian "peace ship," as it was called. I felt strongly that our government shouldn't damage a country by supporting the Contras and sending the CIA to blow up oil tanks.

You were also politically active the following year.

Pauling: Like my friend, George Wald, (Harvard Nobel laureate in biology), who was also on the "peace ship," I also encouraged fellow scientists to campaign against the start of an arms race in space.

In a final effort to get a direct, enlightening answer to a non-hypothetical question, I asked: When you brought the libel suit against

Buckley and *National Review* they listed over 90 organizations you had supported and you didn't deny such support did you?

Pauling: I remember one was "Save the Redwoods."

More than a quarter of a century after the libel trial, I asked Pauling his reaction to the breakup of Communism in the Soviet Union and eastern Europe.

Pauling: In my book, *No More War*! I said the Communist system needs to change, taking on the good aspects of the capitalist system. It would have been better if the Soviet Union hadn't broken up just the way it did. I thought it would have been a more gradual change for the better.

I had failed to talk Pauling into expressing his political views unambiguously, but he is on record as having denied he is or ever has been a Communist. And *Life* magazine conceded "he occasionally criticized some aspects of Soviet tyranny." Even so, in 1970 the Soviet government awarded him the Lenin peace prize. While honored abroad for his peace mission, he lost some friends at home and fought an almost lone battle with the medical community.

In the early seventies he aroused psychiatrists by invading their territory. Seeking a cure for mental illness, Pauling found a link between mental retardation and body chemistry and recommended improved nutrition to help schizophrenics. Scorned by the experts at the time, many of them now admit he's on to something. He also suggested vitamin C as effective cold and cancer therapy.

Nothing has been more controversial than his vitamin C crusade. It was sparked by a chance meeting with biochemist Irwin Stone. In 1966 he heard Pauling, then 65, who had survived Bright's disease, express the wish to live another 20 years to see what would be discovered. Stone promised him another 50 years in good health *if* he took large daily doses of vitamin C, and gave Pauling four research papers indicating its therapeutic value. Skeptical but curious Pauling and his wife became guinea pigs, taking three grams of the vitamin every day. For some three years they were virtually cold free. Pauling published his *Vitamin C and the Common Cold* (W.H. Freeman) in 1970. It also detailed the results of 14 controlled trials, all of which showed the vitamin had some protective effect against common colds. He advised readers to take one or two grams of vitamin C daily for cold prevention, and if a cold occurred to up the daily dose to four grams until the symptoms disappeared.

The book and its author were generally ridiculed by the medical community.

But Pauling defied his critics, stressing that some even misinterpreted their own positive test results. He persisted in his advocacy of vitamin C, concluding, "The evidence is overwhelming that an increased intake of ascorbic acid provides a significant amount of protection against the common cold. The evidence also is strong that it provides some protection against other diseases."[9] And that included cancer.

Speaking at the dedication of a new cancer research center at the University of Chicago in November 1971, Pauling predicted, based on research material, that large doses of vitamin C would reduce by ten percent both cancer deaths and incidence of the disease.

Scottish surgeon, Ewan Cameron, took him up on it and at Pauling's suggestion put his terminal cancer patients on a daily regimen of ten grams of vitamin C. Dr. Cameron continued this treatment at his Vale of Leven Hospital, Loch Lomonside, Scotland, for several years, and undertook a controlled study.

Said Pauling, "All of (Cameron's) 100 cancer patients on vitamin C lived longer than expected—with 90 percent surviving three times longer and 10 percent surviving 20 times longer than doctors thought they would. The average survival time for the treated group was four times longer than that of the control group—209.6 days to 50.4 days. Only three of the 1,000 who didn't receive vitamin C lived as long as a year after they were pronounced terminal. But 16 of the treated group were still alive after $5\frac{1}{2}$ years."

Neither the figures nor Pauling himself impressed Dr. Irvine Page, editor of *Modern Medicine*. Where Buckley had questioned Pauling's loyalty, Page impugned both his professional integrity and his intelligence. No doubt prompted if not incited by his readers, Page ridiculed Pauling as "a tragic example of self-deception . . . who proposed and exploited the use of huge doses of vitamin C for the common cold," and "followed the age-old scheme of claiming a great medical achievement and then demanding that his critics prove him wrong." In doing this, Page accused Pauling of following "the same logic long used by cancer quacks in the sale of useless 'cancer cures.' "

Then Page speculated on what drove this Nobel prizewinning scientist to such "quackery": "When even responsible investigators used shady tactics to promote their 'discoveries' it's no wonder that the public loses confidence in the scientific establishment ... Examples of dishonesty appear to be increasing for multiple motives, such as the desire to receive a grant, political power, or attention."

Pauling had been able to get away with it so far, Page concluded, "if only because of his celebrity status and personal persuasiveness." Page intimated that Pauling was a self-deluded fraud bent on self-aggrandizement.

Six months later, on July 1, 1976, in an action virtually unprecedented in medicine, Dr. Irvine Page completely retracted his criticisms of Pauling.[10] "I do not believe that he acted dishonestly," Page wrote. He also regretted "the unjustified use of the perjorative words 'selfdeception' and 'exploited' in connection with Dr. Pauling."[11]

Both the attack and abject apology got little publicity outside medical circles and isn't even mentioned in Serafini's biography of Pauling.

Shortly after the Page apology, Pauling had to shore up the dwindling funds for his Institute of Science and Medicine, so he advertised in *The Wall Street Journal*: "For Sale—One Thousand Mice With Malignant Cancer—$130 Each."

Pauling: I have often astounded myself. This must have been such an occasion.

Did the gimmick, or act of desperation, work? Were there *any* takers?

Pauling: No. (He chuckled.) One might say the National Cancer Institute. After that advertisement was published we got enough money in gifts to carry out an experiment on cancer in mice and vitamin C. The National Cancer Institute had turned us down for grants five years running, but a few years after that mice ad, they gave us a couple of hundred thousand dollars for us to carry out another mouse study with a different kind of cancer. Both experiments were positive and were published. The National Cancer Institute had said they wouldn't do anything about vitamin C until we carried out animal studies. Then we carried them out, but they didn't do anything. The director told me that if some doctors were to apply for a grant to do follow-up studies they'd consider it seriously. But I couldn't get doctors in medical schools or hospitals to apply for a grant. Probably with the ordinary physician or the medical school professor, he or she wants to avoid the stigma of being associated with quackery. They're afraid to use vitamin C.[12]

But he had more than mice on his mind in those days. Mrs. Pauling, his "second conscience," encouraged him to join his friend, George Wald, a Nobel laureate in biology, in deploring the choice of Milton Friedman for the 1976 Nobel prize for economics.

They wrote that this was "just after the assassination in Washington of the young Chilean economist, Orlando Letelier, former director of the loan division of the International Development Bank, later Chilean Ambassador to the United States from the Allende Government and still later its Minister of Foreign Affairs. On August 28, Letelier published in *The Nation* a critique of the way the Chilean economy is managed under its present dictator, General Pinochet. Perhaps in retaliation for that article, the Chilean Government withdrew Letelier's citizenship just two weeks before his assassination. In that article Letelier characterized Milton Friedman as 'the intellectual architect and unofficial adviser for the team of economists now running the Chilean economy.' Friedman had earlier stated, 'In spite of my profound disagreement with the authoritarian political system of Chile, I do not consider it as evil for an economist to render technical advice to the Chilean Government, any more than I would regard it as evil for a physician to give technical medical advise to help end a medical plague.'

Yes, indeed, a proper physician would treat a wounded criminal; but he must not help him to commit the crime. It should be understood that the government's operations, apart from blatant violations of human rights and the torture of political prisoners, include the suppression of political parties, the destruction of labor unions and church organizations, and the terror maintained at home and abroad by the notorious DINA, the secret police. These are the circumstances within which Milton Friedman performs his 'technical' functions. Presumably the Nobel committee in awarding him the prize took a similarly 'technical' attitude. The Chilean people and the friends of Chilean democracy cannot afford that degree of abstraction.[13]

Pauling and Wald were joined in their protest by a joint letter in the same issue from David Baltimore and S.E. Luria, both Nobel prizewinners for medical breakthroughs.

The Paulings were married for 58 years, until Ava Helen died from stomach cancer in December 1981. She was 77. A former Pauling colleague, Dr. Arthur Robinson, feared "the large doses of vitamin C she had taken might have hastened the growth of cancer." He told Dr. John Grauerholz in a 1984 interview. "She was bathing her stomach with an enormous amount of mutagenic material [material capable of causing change or of inducing genetic mutation] for ten years. I don't know if that's why she got it [the cancer]; there are no statistics . . . but that's the sort of thing I would worry about in the long-term effect."[14]

Pauling has no such worries and quotes statistics to confirm his view that her life may have been prolonged by the vitamin C regime.

Pauling: We began by taking 3 grams a day from about 1966, and then increased it to 10, which she took for the rest of her life. When her cancer was discovered in 1975, she had surgery, a partial gastrectomy. A little more than five years later there was evidence it had returned. An exploratory operation was carried out and it was found that nothing could be done.

Did the surgeon advise her to stop taking vitamin C?

Pauling: Oh, no. He was converted to being a believer in it. A pretty small fraction of patients with a partial gastrectomy because of stomach cancer survive five years. Probably the vitamin C was responsible for her having most of an additional five years of life. She was active until about a month before her death.

Dr. Ewan Cameron also contracted cancer.

Pauling: He died a couple of years ago at about 73. He suffered an injury to his spine from a fall, and then developed cancer of the spine.

Had he been taking large doses of vitamin C?

Pauling: He was a pretty conservative fellow. I tried several times to get him to take it, and he said he was saving it until he needed it. So he wasn't taking it until some time after he suffered the spinal injury [and after he had contracted cancer]. But it was pretty late then.

How do you respond to those who say taking vitamin C didn't prevent you from getting prostate cancer?

Pauling: Perhaps it put it off by 20 years.

I asked Pauling if he still took 10 grams of vitamin C daily and he said he'd increased his intake.

Pauling: For a while I took 12. For several years I've been taking 18 grams. Sometimes I think I'm getting a cold and by taking a good lot of vitamin C I don't have any severe symptoms. I haven't had a severe cold for years.

Despite the cancer deaths of his wife and Cameron and his own prostate cancer, Pauling became more optimistic than ever about the therapeutic value of vitamin C.

Pauling: It pleases me to think I'm making some contribution to solving the cancer problem which causes so much suffering. Though cancer researchers are antagonistic to it, I think a great decrease in mortality from cancer can be achieved in a few years through improved nutrition and especially use of vitamin C. I started out several years ago saying cancer could be controlled to the extent of 10 percent—10 percent fewer deaths with use of vitamin C. And I tend to get up to 75 percent now, as more evidence has been gathered.

The vast majority of the medical establishment still reject Pauling's claims for vitamin C. The Mayo Clinic's randomized double-blind study with 60 patients suffering from advanced cancer found no evidence that daily doses of 10 grams of vitamin C helped them. Pauling responded that the Mayo Clinic had ignored his strong recommendation to use only patients who had not received chemotherapy. He concluded that the Mayo Clinic study was valuable, however, because it verifies that vitamin C treatment is far less effective for patients with immune systems "damaged by courses of chemotherapy."[15]

Some five years later, in 1985, the Mayo Clinic's Dr. Charles Moertel published in *The New England Journal of Medicine* the results of another study he had conducted, titled, "High-Dose Vitamin C Versus Placebo in the Treatment of Patients With Advanced Cancer Who Have Had No Prior Chemotherapy." Moertel wrote: "Patients lived just as long on sugar pills as on high-doses of vitamin C. Vitamin C performed no better than a dummy medicine."

Pauling frequently found flaws in such pessimistic studies.

Pauling: Dr. Cameron's patients received high-dose vitamin C from the time when they began to take it until they died or until the present time, some of them as much as 12 years. On the other hand, the Mayo Clinic patients received high-dose vitamin C for only a short time, median 2.5 months. Moreover, none of the Mayo Clinic patients died while receiving vitamin C. Their deaths occurred only after the vitamin C had been taken away from them.

I'm used to hearing doctors saying vitamin C has no value for the common cold, or any other disease except scurvy. And of them rejecting my arguments, sometimes saying that I was never trained in the field of nutrition. What I think is the important thing is how much knowledge one has about a subject. (As for its use in fighting colds) there's been a lot of work since my books were published. There have been 32 trials with human beings, in the prophylactic value of vitamin C against colds in which the vitamin was compared with a placebo. In 31 the vitamin C subjects usually had a third less illness than the placebo subjects. In the 32nd trial there was no difference Pauling suggested something may have gone awry.

So there were 31 positive and 1 negative.

Pauling: No, there were 31 positive and 1 equivocal. What was equivocal was it involved twins living together. One of the twins was to get the placebo and the other vitamin C. They had the same amount of illness. Then the same investigators did a similar study with the twins living apart. And they had a good result, in that the vitamin C twin had much less illness than the placebo twin. You see, if they were living together they could get their pills confused, or even purposely take the other's tablets.

When a student asked Pauling: "How do you get good ideas?" he replied, "You have a lot of ideas and you throw out the bad ones."

A former student, William Lipscomb, himself a 1976 Nobel prize winner for chemistry, recalls how Pauling was lecturing on his principle
that atoms and molecules should be neutral and a student asked how
he came to that conclusion. "Pauling looked him straight in the eye
and said, 'I made it up.' Which I think is an illustration of how sometimes discoveries are made. They occur. And he tried it. It worked.
It's still one of the most valid laws of chemistry."[16]

What Pauling calls "made up" less playful scientists call making educated guesses or intuition. How does he do it?

Pauling: I think about a problem all the time wherever I am: in bed,
going for a walk, traveling. Then I work at a calculator or a computer
for maybe a week. If I wasn't getting anywhere I'd still think about
it from time to time, especially at night waiting to sleep. It might be
years later that my unconscious, having examined everything I'd
thought about related to the problem, might provide the answer.

Have you any interest in dreams as a source for new ideas?

Pauling: No, I don't think so.

The German chemist, Friedrich Kekule, discovered the closed carbon
ring of the benzene molecule in his dream.

Pauling: There's been a lot of discussion about that. It was pointed
out that it was (about) 25 years after his discovery when he mentioned it had occurred to him in a dream. So, after 25 years his memory may have been faulty.

Have you had any unfulfilled ambitions?

Pauling: I still try to make discoveries. I continue to be interested in
orthomolecular medicine and treatment of diseases especially by vitamin C.* But I've been spending almost all my time on three problems in physics. One is the nature of the so-called icosahevial quasi
crystals. They are crystals of alloys said to have fivefold axis of symmetry, which the crystallographers say don't exist. It's a question of
whether they are single crystals with icosahevial symmetry or are
twins of cubic crystals, as I believe. I'm still trying to get physicists

---

* Pauling coined the word "orthomolecular," to mean the right molecules in the right
  concentration, and suggested that orthomolecular medicine would prevent and treat
  disease with substances normally found in the human body.

in this field to pay attention to my arguments. When you called I was in the middle of a calculation on that. I'm attempting to account for the experimental observations with a different hypothesis.

I hope I haven't spoiled it.

Pauling: No, nothing bothers me in that respect. The second problem I'm working on, and I published a paper on it last year, is superconductivity. The third problem I've continued to work on since 1965 and was thinking about somewhat earlier today is the structure of the atomic nucleus. I've published about 20 papers, but none of the nuclear physicists is willing to pay any attention to it. But I continue to work in that field.

What's your approach?

Pauling: I can illustrate my attack in this way. We have, since 1927, people who make quantum mechanical calculations about molecules, molecular structures. Sometimes they get interesting results. A far more powerful method of thinking about chemical compounds is chemical structure theory. You can make much more progress with the empirical theory based upon a tremendous number of experimental facts than you can by making quantum mechanical calculations. No chemist in his laboratory working on organic, inorganic, or biochemical compounds makes any use of the quantum mechanical theory except in its qualitative aspects. The structure theory was developed from about 1865 on, but has been refined somewhat in the 20th century, especially my paper in the 1930s. And, of course, except for these refinements, it was developed before quantum mechanics was discovered. In the case of nuclear structure there wasn't any empirical information to speak of until after quantum mechanics. The result is that the nuclear physicists just continue to try to solve this Schrodinger equation for a nucleus. Essentially, nobody but me has been striving to develop the sort of structural theory of atomic nuclei compatible with quantum mechanics, of course—but analagous to chemical structure theory. I'm the only person who works hard at that job.

Worldwide?

Pauling: Yes. My first paper published in 1965 has never been referred to in physics journals as far as I know.

Are any of your three goals within reach?

Pauling: I've published half a dozen papers on the icosahedral quasi crystals. I published a paper on superconductivity 20 years ago and another a few years ago. So I've made progress. I've published 20 papers on nuclear structure.

Whom do you support in the genetic engineering controversy?

Pauling: I'm not on either side. I don't work in the field. I don't have knowledge and experience enough for me to make reliable statements about it. I do advocate caution, but I've avoided making public statements about it. I tend a little bit toward the government-control side.

The first major biography of Pauling, by Anthony Serafini, pictures him as an infuriating, abrasive, self-absorbed publicity hound, often abrupt and intimidating, apt to disparage others and not above pirating their ideas. In this 1989 biography, Serafini writes: "Another of Pauling's great talents was coming into play and more often—his talent for self-promotion. For some time he had been working on a translation of Goudsmit's famous book on the structure of line spectra. However, numerous scientists at Caltech, many of them long-time acquaintances of Pauling's, seemed to believe he had actually written part of *The Structure of Line Spectra*, and that he had contributed substantially to the ideas in the book. (But) it was Goudsmit's work alone, however."

When I read this to Pauling, he said, "Serafini said everything in the book was taken from the famous book by Goudsmit. Well, there isn't any famous book by Goudsmit: he never wrote a book before this one. He was a very interesting fellow.* I became acquainted with him in 1927 and helped him in his study of spectra and he suggested we write a book on the vector model together, which we did, in 1930. And the archivist at the Pauling Archives at Oregon State University wrote me saying they have a letter from Goudsmit to me saying that since I had been responsible for most of the material in the book and had written most of the book my name should come first of the two authors."

---

* Samuel Goudsmit, chief scientific adviser to the Alsos Mission in World War II, helped American Intelligence agents capture and interrogate German scientists they thought might be trying to produce an atomic bomb. Among them were Heisenberg, Hahn, and von Laue. Goudsmit had invited Pauling to join him on this adventure. Pauling, otherwise engaged, also didn't think his German was up to it. Goudsmit is noted as codiscoverer of electron spin and as a man with a quip for most occasions. His father managed a toilet-seat factory until, said Goudsmit, "the bottom dropped out of the business." In 1978 at 76, he died in his car of a heart attack.

Are there other biographies in the works?

Pauling: There's been only one other, by Mrs. White for teenagers.[17] I don't know how far along other biographies are.

Serafini wrote that after Pauling gave a talk at Smith College, Professor Marjorie Senechal asked a question and he responded by saying "Where's my coat?" to his wife, and making a quick exit. His questioner thought she deserved better than that. I asked Pauling for his response.

Pauling: Perhaps I was just walking away and didn't stop. I don't know what the question was or what she wanted to talk about.

Is it likely that if you don't like a question you simply don't answer it?

Pauling: No, I don't think so. (He laughed.) In general, I'm quite polite in my behavior.

Fellow scientist, Dr. Dorothy Wrinch is quoted in Serafini's book as saying, "Pauling is bright and quick and merciless in repartee when he likes, and I think people are just afraid of him. It takes poor Dorothy to point out where he is wrong."

Pauling: I've known some people who were not above trying to embarrass others. I've heard stories about Oppenheimer, for example, embarrassing Teller. But it's quite foreign to my nature.

Perhaps you didn't hear Professor Senechal's question or were in a hurry, and people take offense when none's intended.

Pauling: Sometimes after I've given a lecture there's a big crowd, especially young people, wanting my autograph. It may be that I'm getting tired. Usually I sign, but if it gets too burdensome I say I'm not able to give any more autographs.

Since Pauling wouldn't hand over or quote from letters by former colleagues disputing Serafini's charges, I contacted one willing to talk. Alexander Rich had worked with Pauling from 1949 to 1954 and had been in touch with him quite steadily ever since. He also knew Samuel Goudsmit. Now a biophysics professor at M.I.T., Rich said: "Serafini's biography was outrageous, sloppy, and unscholarly. And the fellow made so many errors. It's a thoroughly discredited piece of work. The charge that Pauling contributed very little to the Goudsmit book is completely false. Goudsmit had done this thesis and Linus rewrote it extensively and published the book. I talked to Sam Goudsmit before he died (in 1978), and there was no question about the extent to which Linus modified things. It's a matter of simple record. If one goes and looks at the thesis, then at the book it's quite clear. But Serafini didn't do that or couldn't do that, or whatever. I looked up several things where I was personally involved and I could see he had made mistakes."

A *New York Times* headline for October 19, 1979, reads, "Proving Professor Wrong Led to Nobel Prize for Lipscomb." And the following news account reads: "When William N. Lipscomb was a graduate student at the California Institute of Technology, he heard a professor propose a theory of chemical bonding in certain boron compounds. Something about the theory did not seem right, and he set out to prove his professor wrong. In so doing, he won the 1976 Nobel Prize in chemistry. The professor was Linus Pauling, a Nobel laureate and a strong influence on the career of Dr. Lipscomb. In fact, he used a technique taught him by Dr. Pauling to identify and study the structure of boranes, the complex compounds combining the element boron with hydrogen."

Professor Alexander Rich set the record straight, telling me on March 1, 1993: "Lipscomb didn't prove Linus wrong. Linus described what was going on in boron chemistry as best it was known at the time, and not much was known. What Lipscomb did was go into an area that had been unexplored and made a number of discoveries."

Many medical experts still reject Pauling's claims for vitamin C, but Professor Rich says: "I've read a great deal recently that lends support to the importance of vitamin C. It is believed that one mechanism producing cancer is damage to DNA, and recently a great deal of work's been done on the role of antioxidants, things that prevent oxidation, of which vitamin C is important. Oxidation changes material chemically. Vitamin E and vitamin A derivative are also important in preventing damage to DNA. It's also important in preventing oxidation of a fat in the blood that leads to arterial sclerosis."

Does this validate Pauling's belief that an enzyme, by weakening the "cement" between cells, allows cancer to spread? And that vitamin C inhibits this process? Is the vitamin, in fact, an effective cancer therapy?

"I feel," said Rich, "that the correlation of antioxidation protecting against oxidative damage and the fact that oxidative damage is an important part of tumors makes a plausible case for it."

During one conversation with Pauling I hoped he would entertain a wide range of questions and I wasn't disappointed. I knew he didn't believe in God, but how about life after death?

Pauling: Don't believe in it at all, not even the possibility.

Is intelligent life on other planets plausible?

Pauling: Surely, very highly. In the solar system there are so many stars and so many planets and the process and the origin of life is a pretty straightforward one that will occur when conditions are right, when the chemical composition, temperature and other factors are right. It's likely to be based on carbon, nitrogen, oxygen, hydrogen.

What do you think of UFOs or flying saucers?

Pauling: After looking over the evidence I'm thoroughly skeptical.

Does hypnosis interest you?

Pauling: No.

The physicist, Richard Feynman, told me he investigated hallucinations by spending time in one of John Lilly's isolation tanks. Would something like that interest you on the offchance it might give you bright ideas?

Pauling: I'd read about Lilly's work, trying to communicate with porpoises. But I didn't know about the isolation tanks. When I was about 28, lecturing every year at Berkeley for a month or two in the spring, I got a number of books out of the library on occult phenomena. And Bateman and Soal in England sent me a copy of their book. (*Modern Experiments in Telepathy*, S.G. Soal and F. Bateman, Yale, 1954.) I read all these books and decided there wasn't any evidence in anything I read that seemed likely enough to be significant enough to justify spending any time on the matter. So it was about 60 years ago that I lost interest in phenomena of that sort. That didn't mean I didn't continue to have original ideas.

You knew Paul Dirac.

Pauling: In 1926 or 7 I met him in Gottingen and several times since, once in Florida a year or two before his death. He was a man of very few words. I didn't discuss scientific matters with him because we were interested in different levels of science. The same way with Einstein. My wife and I used to spend an hour with him every time we were in Princeton. We also didn't talk about science because he was interested in this very fundamental problem (unified field theory), and I was more interested in applying quantum mechanics to chemical problems. I never discussed philosophy with Bertrand Russell the times we met because our levels of interest were different.

After Einstein died, Dirac was called the greatest living physicist.

Pauling: I don't like to make statements of that sort, but I can accept that all right.

Did you know Dirac was religious? His wife told me he believed in Jesus Christ.

Pauling: In what respect? Some say there was never any such person in existence.

I presume she meant as God.

Pauling: I don't think she's reliable, any more than Eugene Wigner is. He is emotional about nuclear weapons and questions about the Soviet Union, in the same way that Teller is. I was involved, as I said, in that formal debate with Teller on television with a stopwatch, each of us having two minutes and then five minutes and then three minutes, and so on. I was involved in a debate with (Eugene) Wigner, [Nobel prizewinning physicist] too, in the pages of a journal. In each case I felt that the person, Hungarian, with that sort of experience involving the Soviet Union was governed to such an extent by his emotional feelings and convictions that he was no longer rational when it came to discussing problems of that sort. Rational enough on scientific matters, of course. Both Wigner and Teller are very able scientists. I think they both started out as chemical engineers, the way I started, too. But when it came to political matters the emotional factor overcame them. In the same way, Mrs. Dirac might be speaking from an emotional basis when she said he had believed in Christ, by saying something she would like to believe about Dirac.

She also said she believed in telepathy—when she was thinking of her daughter, the daughter phoned, that sort of thing.

Pauling: I'm not surprised.

She said that when she discussed her belief in telepathy with Dirac he was skeptical. What's surprising is that he never told her something he told me: that in his early days at Cambridge University he made a number of telepathic experiments and concluded telepathy did not occur. But he never told his wife.

Pauling: He may have been cautious. [He chuckled.] There's no use in upsetting your wife, you know.

Do you work every day?

Pauling: Essentially, since my wife died in 1981. I don't have anything else to do.

How long do you sleep and what's your typical day?

Pauling: Between six and ten hours. I get up at 7:20 and go to bed at 10. I said 7:20, but there's a standard deviation of 20 minutes. I read the literature and do a pretty full day of work, scientific and medicine, as a rule.

Your resonance theory was rejected by the Soviet Union because it wasn't in strict accordance with Soviet ideology. Do they accept it now?

Pauling: Yes. There was a period of about five years, from 1950 to 1955, before they got around to reversing the previous action. When Stalin's influence had faded.

Did Robert Oppenheimer get a fair deal?

Pauling: About the investigation? Not at all. I spoke about that at the time. The authorities knew all about him before he began his work at Los Alamos, and didn't learn anything new. I think it was a matter of jealousy on the part of such people as Strauss and Teller.

Were you an admirer of human rights activist Sakharov?

Pauling: Yes. And I signed several petitions supporting him. I can't say I agreed with everything he said and did. We were somewhat alike. I am a physicist in part. We've both been in trouble with our governments. One difference between us is that he worked in nuclear arms development. I never have.

Have you changed your views much since your youth?

Pauling: In one way. When I was younger I accepted the dictum expressed over and over again that even people who know quite a lot about science aren't capable of discussing public affairs. It took me 45 to 50 years to decide that's wrong, that scientists have as much right to discuss public affairs, if they're interested in them and have background knowledge, as anyone else.

Although some have called you dangerously naive about Russia and Communism and, referring to your criticism of President Kennedy's action during the Cuban missile crisis, say your great defect is blindness to alternatives, didn't you realize the alternative to stopping nuclear weapons getting to Cuba? Are those fair comments?

Pauling: I don't know whether fairness enters into it. I think they're probably incorrect if they suggest a lack of knowledge or understanding on my part.

What makes you angry?

Pauling: Not very much. I'm pretty even tempered.

In a published interview some years ago you seemed angry through the whole thing.

Pauling: A woman with *Cosmopolitan*. I got irritated with her right at the start because she'd come to interview me about the common cold and hadn't read my book *Vitamin C and the Common Cold*. And she wanted me to go through it section by section. [He chuckled.] To do her work for her, instead of having read the material and formulated questions to clarify points. She had a disagreeable manner, as well. It's really very rarely that I get irritated by any person.

What makes you laugh?

Pauling: Lots of things. The comic strips, for example. I don't like Orphan Annie. There are many I do like. Doonesbury, Gordo, and Fred Basset. The great Chaplin. I like him very much.

You seem very vigorous.

Pauling: I'm about as vigorous as I ever was, but physically I'm not quite up to what I used to be. I used to be able to squat down from a standing position holding one leg out in front of me until I was sitting on the ground, and then rise without ever letting that leg touch the ground—rise on one leg. Well, I have trouble squatting down with two legs now. The only problem I have isn't very serious: rather poor teeth. I still have most them, but I have to go to the dentist occasionally. That's probably the result of growing up in Willamette Valley, during most of my early years, where people drank melted snow water containing minerals or fluorides. It was also an area low in iodine.

Any big disappointments in your career?

Pauling: I've been so pleased with things I've done that I haven't worried about those I failed to do.

My last conversation with Pauling was on March 14, 1993.

Now you're 92, Dr. Pauling, how do you feel?

Pauling: Pretty well. I strained my knee recently, but it doesn't bother me very much.

Like Einstein, you seem to prefer living alone.

Pauling: In a sense. I'd prefer to be continuing to live with my wife. Since her death I've preferred being by myself.

You still wake to the sound of waves?

Pauling: Whenever I wake up I hear the ocean. When the waves are heavy there's a lot of noise of the rocks being rattled against one another.

What time did you get up?

Pauling: I woke about five, but didn't get up for two or three hours. I read in bed, a detective story, and some popular science.

No newspapers?

Pauling: When I'm in Palo Alto, *The San Francisco Chronicle.* Here I just rely on television.

During his years of protest, Pauling had not only joined demonstrations against nuclear tests, but had frequently spoken out against the Korean and Vietnam wars. Yet he said he was not opposed to all wars. I asked for an explanation.

Pauling: Whenever we make a decision it should be on the basis of the estimated amount of human suffering associated with the alternatives. War is one of the principal causes of human suffering. Consequently, we should prevent wars as not compatible with ethical principles of humanity. But there may be circumstances in which an oppressive and dictatorial government causes so much suffering, a war of liberation might be justified on the basis of my ethical principle.

Like World War II.

Pauling: Oh, yes. Well, I received the Presidential Medal for Merit from President Truman for contributions I made to the war effort during World War II. Like most scientists, even those like me who are very strongly opposed to war, I felt Hitler was such a menace to the world that Nazism had to be stopped.

How about the Persian Gulf War, Desert Storm?

Pauling: You may have seen my two advertisements I paid for—one in the *New York Times*—saying, "Stop the trend toward war. Rely on negotiations. Don't attack until other pressures have been applied." And another one after the attack had begun, killing hundreds of thousands of Iraqis and perhaps 50 Americans. I published another ad in a Washington D.C. paper saying, "Stop the War Now! Go back to resorting to diplomacy and other pressures."

You think Saddam Hussein would have been susceptible to negotiations without force?

Pauling: If the pressures had been applied in the proper way.

How about Yugoslavia today?

Pauling: It's a miserable situation. It needs to be solved without the killing and suffering associated with war. I just don't know how to solve it. It's existed for a thousand years or more: these different ethnic groups with somewhat different languages and religions attacking one another in a way which is quite incompatible with religious

principles. I'm strongly opposed to how the situation is being handled. On the other hand I don't know what the best way is.

I've been in Slovenia, Croatia, and Serbia, but I'm not an authority on the situation. I feel strongly that the leaders of the nations
of the world are derelict in their duty in that they have not striven
hard enough to prevent the suffering in that region.

When we spoke last you were working on three physics problems.

Pauling: I still work on them, but I move from one to another. Right
now I'm working on a paper about vitamin C and cancer. I work at
it every day, go to bed pretty early and get up pretty early.

You'll be alone today, I take it.

Pauling: I'm alone every day.

You weren't a few days ago.

February 28, 1993, was his 92nd birthday. About a third of his
extended family, 12 or so, turned up. His physician son, Linus Jr. and
wife gave Pauling a diploma for the occasion.

Pauling: They thought the one thing I needed was a diploma. [He
chuckled.] It says, "Stephanie and Linus Pauling Jr. present to Linus
Pauling loving greetings on his 92nd birthday. Happy birthday, Pop!
Best wishes and love. Stephanie, Linus, on this Sunday, 28th of
February, 1993." Then there are four seals which are in fact American
postage stamps about four different minerals. I didn't know the Postal
Department had issued those, so I was glad to see them.

And now you're back to work.

Pauling: Yes. Most of it is on vitamin C in relation to disease, cancer,
and the sort of cancer that AIDS patients have. There's a disease that
causes paralysis and many deaths in India and South America. And
Dr. Dunham in our Institute has shown with mice that the paralysis
is prevented with high doses of vitamin C.

How will you spend the rest of today?

Pauling: I'm working on a paper on vitamin C and other vitamins
and cancer. I have some calculations to do, biostatistics. Then I'll send
it to *The Journal of Orthomolecular Medicine*.

Someone had recently mentioned that the balcony outside his cabin was in danger of collapsing. I checked this with a Pauling daughter-in-law who said it was safe enough though it may have some dry rot.

When you stand on your balcony how far are you from the ocean?

Pauling: About 140 feet at an angle of 45 degrees. That's 100 feet horizontally and 100 feet vertically. (To him it's an isosceles triangle. To me it's a risky proposition.)

If it collapsed, would you fall in the ocean?

Pauling: No, I'd fall on the ground and if I weren't stopped by brush, trees, (like Hemingway he skipped articles, maybe to save energy.) I'd roll down a slant about halfway and then fall off a cliff about 50 feet high.

My God! (But he wasn't concerned, and obviously cherished his solitude.)

Linus Pauling's stubborn, crusading spirit took him around the world to recruit others to the cause.

During a fervent speech against nuclear weapons, Pauling said, "Every human being should have this feeling of exaltation, this feeling of deep happiness that comes from believing that you are doing something worthwhile."[18]

He complained that his wife held him to almost impossibly high standards. But he didn't disappoint her. "What I have admired and liked in him," she said in a television interview, "is the fact that he has worked with such passion and eagerness and truth in science, regardless of the awards and honors that he has received. And the help that he has been to so many people."[19]

John Dowling calls him "a giant of his times . . . Few of us have his brilliance or integrity or courage, but his words and actions are encouragement enough to help us continue the fight against the arms race."[20]

Pretty fair assessments.

Chapter Two.
Richard Feynman

Feynman is "a second Dirac, only this time human."
—Eugene Wigner
"The first principle is that you must not fool yourself,
and you're the easiest person to fool."
—Richard Feynman
An admirer saw him as "not a person at all, but a more advanced life form
pretending to be human to spare our feelings." Others called him selfish,
brash, irresponsible, and arrogant.
When Omni magazine rated Feynman the world's smartest man, his otherwise
adoring mother said, "If he's the world's smartest man, God help us!"

Richard Feynman beat bongo drums for hours, got drunk, and picked locks guarding secret documents, while helping to unleash a new terror on the earth—the atomic bomb. At the Manhattan Project, as it was called, this hellraiser was considered the brightest young physicist of them all—Einstein's heir apparent. The military commander, General Groves, called him and his colleagues "the largest collection of crackpots ever seen." In a vote, Feynman would doubtless have been crowned crackpot-in-chief.

Feynman became a national celebrity when he appeared on TV holding aloft a small piece of rubber, an O ring seal, of the kind used on the ill-fated space vehicle Challenger. As the camera focused on him, 68-year-old Feynman (who had carefully rehearsed this moment) dipped the rubber seal into a glass of ice water for a few moments. Then took it out. Presto! It had lost its elasticity. So Feynman solved the mystery. His simple demonstration showed that when subjected to freezing temperatures the seal would become brittle. And that caused the Challenger to disintegrate in flight.

A physicist friend, Freeman Dyson, applauded Feynman's feat, especially because "the public saw with their own eyes how science is done, how a great scientist thinks with his hands, how nature gives a clear answer when a scientist asks a clear question." Like Einstein, Feynman searched first for simple answers. Also like Einstein, he said exactly what was on his mind. Asked during World War II if he considered a former secretary suitable for a job in Naval Intelligence, Einstein replied: "No, because she is not intelligent." And Feynman, sitting next to a Danish princess, disputed her remark that they couldn't discuss his subject, physics, because nobody knows anything about it. On the contrary, Feynman replied, that's just what everyone does discuss—things they know nothing about: the weather, social problems, psychology, international finance. Knowing a putdown when she heard it, the princess turned her attention to a less provocative dinner guest.

Unlike Einstein, Feynman viewed almost everyone outside the scientific world with contempt and all organized religions as baloney. He despised shallow thinkers and dabblers in the arts, yet taught himself to sketch for a hobby. He berated philosophers as stupid and poets as sissies, yet he had a poetic streak, and was a minor master of metaphor and imagery. He was tone deaf, and classical music caused him physical pain, but he beat bongo drums for hours, enjoying the percussion effect.

He escaped military service in World War II by fooling the examining psychiatrist into listing him as mentally defective, which left him free to play a more devastating wartime role, building the atom bomb.

Richard Feynman was born in Far Rockaway, Queens, on May 11, 1918. His father, Melville, had emigrated as a child from Minsk, Byelorussia. In America he became manager of a company selling police and mail carrier uniforms. Feynman's mother, Lucille, was the daughter of a man who had emigrated as a child from Poland and was raised in an English orphanage. As an adult he emigrated to the U.S. and became a successful milliner. Lucille's mother suffered from epilepsy and her eldest sister was schizophrenic. Feynman's sister, Joan, nine years his junior, was to follow him into science, with a Ph.D. in solid state physics.

Though not religious, his Jewish parents sent him to Sunday school at a Reform Synagogue apparently on the do-as-I-say-not-as-I-do principle. Again like Einstein he was shocked to find stories from the Bible were a mix of fact and fiction—and quit Sunday school in disgust. He never wavered in his antireligious feelings.

Feynman attended high school in Far Rockaway, but learned his more lasting lessons from his father, who taught him not to kid himself that he knew anything about a thing simply because he could name it. The names of birds, for example, he said, tell you nothing worthwhile about them. You must listen to them, study them, and think about them. Feynman adapted and expanded this advice to consider things from unusual angles, as if, for example, he were a Martian visitor to the earth.

However as will be explained later, on one occasion, not knowing the name of something proved embarrassing.

His math prowess took him to Massachusetts Institute of Technology (MIT), where he continued to practice "Martian thinking." This led him to visualize how we might appear to Martians: as creatures attached by some mysterious force to a ball, half of which is upside down, and is spinning around in space. The ball circles a great glob of gas [the sun] that's burning fuel.

Doing research on sleep he imagined himself as a Martian who never slept, then arrived on earth and saw people who had to lie down and be unconscious for several hours at a time. Feynman the Martian would wonder: What's it like to be unconscious? What happens? Do you suddenly stop thinking of ideas or does your thinking gradually slow down before you become unconscious? And what happens to ideas you had when you were conscious?

His father also taught him to have no respect for authority but to study things for himself from start to finish.

Feynman was judged "socially inept," at MIT, as well as "athletically feeble, miserable in any but a science course, risking laughter every time he pronounced an unfamiliar name, so worried about the other sex that he trembled when he had to take the mail out past girls sitting on the stoop."[1] But he delighted children with conjuring tricks, nonsense talk, and imitations of pompous colleagues. He taught dogs to do tricks, ants to move at his command, and himself to do a comic imitation of a one-man band.

At Princeton he worked on his Ph.D. thesis titled "The Principle of Least Action in Quantum Mechanics," with frequent breaks. One afternoon he studied ants moving to and from a box of sugar he had suspended on a string, hoping to learn how they communicate. He concluded they leave tracks for others to follow. He also learned how to lure them away from his larder with a piece of sugar and that some ants aren't all that bright. One cold winter day a fellow student found Feynman at an open window, stirring a pot of Jell-O and shouting at all who approached to get lost. He was trying to see how Jell-O co-agulates while in motion. And on another occasion, when an argument erupted about the motile techniques of human spermatozoa, Feynman disappeared and soon returned with a sample.

Believing he could partly control his dreams and steer them in the desired direction, he put his theory to a critical test in one dream in which he was eagerly approaching three attractive young women in bathing suits. But his theory took a tumble when the women turned into old men playing violins.

When he graduated in 1942, World War II was waiting for him. Feynman deliberately responded to questions from a military psychiatrist with wacky answers, and was rated mentally defective. This kept him out of the army, leaving him free to join hundreds of other scientists Los Alamos bound. He was just 24 when the director of the Manhattan Project, Robert Oppenheimer, called the "mental defective" the most brilliant young physicist there and made him a group leader. Not bad for a young man who couldn't tell left from right without first checking the mole on his left hand.

But he knew a looming disaster when he saw one. Uranium was being purified at the Oak Ridge Plant in Tennessee to use in the atomic bomb. Afraid the secret would get out, General Groves did not warn the workers of the danger. Consequently such massive amounts of volatile uranium were piled in one place that the plant "risked blowing itself up altogether." Feynman was sent to the rescue and saved 85,000 workers from the "inevitable" explosion.[2]

The first atomic bomb test, code-named "Trinity," was on July 15, 1945. As Oppenheimer watched their handiwork explode, and thought of himself with anguish as a potential world destroyer, Feynman sat on the back of a jeep beating a triumphant tattoo on his bongo drums. Many of his antics in those days were to entertain his young wife, Arline. She had been his high school sweetheart whom he had married knowing she was dying from tuberculosis. Every week he traveled from Los Alamos to the Albuquerque hospital where she was slowly growing weaker. She died in 1946. On the anniversary of her death, he wrote her a love letter, discovered after his death, saying "other women all seem ashes" and ending, "P.S. Please excuse my not mailing this—but I don't know your new address."[3]

He remained single for seven years but between marriages took on another research project in Las Vegas: learning to distinguish showgirls from hookers, and how to pick up the former. But his work always came first. In fact he turned down an offer to succeed Fermi at the University of Chicago for fear the increased salary would tempt him to take an expensive mistress who'd divert him from his first love—science.

His clowning irritated Niels Bohr, especially at a colloquium where Feynman heckled the speaker. Though no kill-joy, Bohr criticized Feynman for playing the fool, and embarrassed him at a 1948 conference chaired by Oppenheimer and attended by other big guns of physics, Bethe, Bohr, Dirac, Fermi, Rabi, Teller, Wheeler, and Von Neumann. Feynman's lack of interest in the names of things proved his downfall. Expressing his latest ideas with chalk-board equations, he was interrupted by Dirac asking if it was unitary. Not knowing the meaning of the word, Feynman craftily suggested Dirac hear him out and then tell him if it were unitary. Moments later, Teller complained that Feynman had forgotten Pauli's exclusionary theory. Feynman worked his way out of that only to be told by Bohr in a devastating putdown that he had forgotten the most important lesson learned from two decades of quantum mechanics. Bohr finally took pity, and the chalk from him, and put him back on track.

Nevertheless, Hans Bethe sensed Feynman's potential and got him a job at Cornell. Feynman taught there until 1951 when, after spending six months in Brazil, he became a professor at Caltech. He stayed there for the rest of his life.

Some students were spellbound by his lectures, others walked out, unable to take the pyrotechnics or intimidated by his demanding standards or provocative style.

His second marriage, in 1952, to Mary Louise Bell, was a disaster, and lasted barely four years. She called him "an uneducated man with a Ph.D.," and complained of his violent temper, noisy bongo drums, and habit of taking work to bed with him. "He begins working calculus problems in his head as soon as he awakes," she said. "He did calculus while driving his car, while sitting in the living room and while lying in bed at night."[4]

After his divorce Feynman had several turbulent affairs, was hounded by the women he had unceremoniously dumped, and blackmailed by a cuckolded husband. He fled to Switzerland in 1958 to get away from it all. There, at 40, he met his third and final wife-to-be, Gweneth Howarth, a 24-year-old Englishwoman working as an au pair. Even his courting tactics were unique. He first hired her as his live-in maid before marrying her in 1960. Their son, Carl, now a computer scientist, was born in 1962. They adopted a daughter, Michelle, in 1986.

He won the 1965 Nobel prize in physics for insight into quantum electrodynamics with important consequences for the physics of elementary particles. He shared it with the Japanese Sin-Itero Tomonaga and fellow American Julian Schwinger.

Victor Weisskopf had admired Feynman at Los Alamos not only for his talent as a problem solver, but because of his charming personality. "To our children, who adored him, he was the funniest man in the world and their favorite adult playmate. . . . Many times after the war I went to Pasadena . . . and asked for his help in understanding this or that physics problem. He always managed to use the right language, leaving out mathematical complications."[5]

In 1985 Feynman published an informal memoir, *Surely You're Joking Mr. Feynman!*, based on conversations with a friend, Ralph Leighton.

His Alix G. Mautner Memorial Lectures were also published that year. Titled *QED* [Quantum Electrodynamics]: *The Strange Theory of Light and Matter*, they forced Feynman to reject his father's advice to accept only ideas that seem reasonable. As he told his audience, "The theory of quantum electrodynamics describes . . . all the phenomena of the physical world except the gravitational effect . . . and radioactive phenomena. [It] describes nature as absurd from the point of view of common sense. And it agrees fully with experiments. So I hope you can accept nature as She is—absurd."[6]

The following year he shot down the idea that there was a fifth force at large in the universe, informing *The Los Angeles Times*, "Such new ideas are always fascinating, because physicists wish to find out how nature works. Any experiment that deviates from expectations according to known laws commands immediate attention because we may find something new. But it is unfortunate that a paper containing within itself its own disproof should have gotten so much publicity. Probably it is a result of the author's over-enthusiasm."[7]

In February 1986, with millions of others, Feynman was watching the televised launch of the space shuttle *Challenger* and saw it explode, killing all aboard. A few days later, he accepted NASA head William Graham's invitation to join the Presidential commission investigating the disaster. Having been one of Feynman's students at Caltech, Graham knew Feynman would not tolerate any whitewash or evasion. His interrogation of space-shuttle engineers illustrates his ability to "cut the crap," as he would have put it. He recalled: "They kept referring to the problem by some complicated name—a 'pressure-induced vorticity oscillating wa-wa,' or something something. I said, 'Oh, you mean a whistle.' 'Yes,' they said, 'it exhibits the characteristics of a whistle.' "[8]

The demonstration of his discovery shown live on TV—that frozen rubber seals caused the Challenger disaster—sustained his almost legendary reputation.

Even measured against other Nobelists, Feynman stood out as unusually bright, original, and challenging. He was an arresting and amusing lecturer, always on the move as if about to run for it; advising students that the only way to succeed in science was to ignore your own feelings and describe the evidence very carefully and honestly; and to be critical of your own theories, remarking with equal fervor both weak and strong points. Integrity was the watchword. "We've learned from experience," he said, "that the truth will come out. Other experimenters will repeat your experiment and find out whether you were wrong or right. Nature's phenomena will agree or they'll disagree with your theory. And, although you may gain temporary fame and excitement, you will not gain a good reputation as a scientist if you haven't tried to be very careful."[9]

He demonstrated his quick wit during a lecture when his remark, that the gravitational force is *damned* weak, was followed by an overhead loudspeaker crashing to the floor. After a pause worthy of Noel Coward, he said, "Weak, but not negligible." He never ceased to be amazed that "everything is made of the same atoms, us, the air we breathe, and the grass we walk on," and that "according to the exclusion principle no two electrons can occupy the same space. This holds true for quantum mechanics and relativity. Yet with quarks the opposite is true: two particles tend to occupy the same space."[10]

He decided that if only one scientific truth could be saved for future generations, presupposing some universal catastrophe, it should be that all things are made of atoms, moving around in perpetual motion, attracting and repelling each other.

In a more poetic vein he wondered about the human brain: "So what is this mind of ours? What are these atoms with consciousness? Last week's potatoes! They now can *remember* what was going on in my mind a year ago—a mind which has long ago been replaced . . . The atoms come into my brain, dance a dance, and then go out—there are always new atoms, but always doing the same dance, remembering what the dance was yesterday."[11]

During his six months of lecturing in Brazil he expressed his creed to a student audience: "Science is a way to teach how something gets to be known, what is not known, to what extent things are known—nothing is known absolutely—how to handle doubt and uncertainty, what the rules of evidence are, how to think about things so that judgments can be made. And how to distinguish truth from fraud, and from show."[12]

He published his last book in 1988, the year he died. Titled *What Do You Care What Other People Think?*, the title came from a comment made by his first wife, Arline. It includes hilarious memories, a moving account of Arline's last days, and details of his successful investigation of the cause of the space shuttle *Challenger* disaster.

But Feynman "didn't write a word of his two books," said his secretary, Helen Tuck. "They were a result of conversations over the years with his friend, Ralph Leighton, who recorded them, sometimes after they took a break from their bongo drum duets. Dr. Feynman did edit the books closely and had great fun working with Ralph Leighton and Ed Hutchings on the proofs."[13]

During a lecture at Caltech, Feynman spoke of his encounters with Robert Wilson, Robert Oppenheimer, Hans Bethe, Enrico Fermi, Klaus Fuchs, John von Neumann, and Niels Bohr:

WILSON

At Princeton, physicist Bob Wilson "said he had been funded to do a job that was a secret, and was going to tell me because he knew that as soon as I knew what he was going to do I would see that I had to go along with it. So he told me the problem of separating different isotopes of uranium ultimately to make the bomb . . . And he said, 'There's a meeting at three o'clock. I'll see you there.' I said, 'It's all right that you told me the secret because I'm not going to tell anybody—but I'm not going to do it.' But I went back to work on my thesis for about three minutes and then I began to pace the floor to think about this thing. The Germans had Hitler, and the possibility that they'd develop the bomb before we did was very much of a fright. So I decided to go to the meeting at three o'clock. By four I already had a desk in a room and was trying to calculate whether this particular method was limited to the total amount of current you can get in an ion beam, and so on. I was working as fast as I could and as hard as I can, whereas the fellows who were building the apparatus to do the experiments right there—the other fellows who had joined—it was like those moving pictures you see where you see a piece of equipment go berup! berup! berup! Every time I'd look up, the thing was getting bigger. And what was happening, of course, is that all the boys had decided to work on this and to stop their own research in science—all science stopped during the war except the little bit that was done at Los Alamos. It was not much science. It was a lot of engineering. And they were robbing their equipment from their research. And all the equipment from different research was being put together to make the new apparatus to do the experiments to try to separate the isotopes of uranium. And I stopped my work, also, for the same reason."

OPPENHEIMER

"I met some very great men at Los Alamos. There were so many of them, it was one of my great experiences in life to have met all those wonderful physicists. Oppenheimer was very patient with everybody. He paid attention to everybody's problem. He worried about my wife who had TB and whether there would be a hospital out there. He was such a wonderful man."

BETHE

"All the big shots by some kind of accident, everybody but Hans Bethe, happened to leave [Los Alamos] at the same time. Weisskopf had to go back to MIT, and Teller was away at a certain moment. And what Bethe needed was somebody to talk to to push his ideas against. So he came in to this little squirt in an office and he starts to argue, to explain this idea. And I say, 'No, you're crazy! It will go like this.' And he says, 'Just a moment.' And he explains how he's not crazy, that I'm crazy. And we keep on going like that. You see, when I think about physics I don't know who I'm talking to, so I say the dopiest things like, 'No, no, that's wrong!' or, 'You're crazy!' But it turned out that's exactly what he needed. And I got a notch up on account of that. And I ended up a group leader with four guys under me, which is underneath Bethe. You see, there were several groups. I had a lot of interesting experiences with Bethe. The first day we came in we had an adding machine which you work with your hands. So he says, 'Let's see, the formula I've been working out involves the pressure squared. The pressure's 48. The square of 48 . . .' I reached for the machine and he says, 'about 2300.' So I plug it out just to find out and he says, "You want to know exactly? It's 2304.' [laughter] So it came out 2304. I say, 'How d'you do that?' He looks at me and says, 'Don't you know how to take squares of numbers near 50? If it's near 50, say 3 below, then its 3 below 25, like 47 squared is 22, and how much is left over is the square of what's the residual. For example, you've 3 lost so you get 9, 2209 from 47 squared.' Very nice. Okay. [laughter] He was very good in arithmetic. So we kept on going and a few moments later we had to take the cube root of $2\frac{1}{2}$. Now to do cube roots there was a little chart and you have some trial numbers you do on the adding machine.

Hoping this takes him a little longer, you see, I open the drawer, I take out the chart, and he says, 'One point three five.' [laughter] So I figure there's some way to take cube root numbers near $2\frac{1}{2}$, but it turns out no. [laughter] So I said, 'How d'you do that?' and he says, 'Well, you see, the logarithm of $2\frac{1}{2}$ is so and so, you divide by 3 to get the cube root of so and so. Now the log of one point three is this, the log of one point four is this. [laughter] I interpolate in between.' I couldn't have divided anything by three, much less . . . So he knew all his arithmetic and that was a challenge to me. I kept practising. We used to have little contests. Every time we'd have to calculate anything, we'd rush to the answer, he and I, and I'd win now and then. After several years I'd be able to do it, you know, get in there maybe one out of four. Because you'd notice something funny about a number, like if you have to multiply 174 by 140, for example, you notice that's like 173 by 141, which is like [speaking rapidly] the square root of three times the square root of two which is the square root of six which is 245. [laughter and applause] But you have to know the numbers, you see. Each guy would notice a different way, and we had lots of fun."

FERMI

"The first time Fermi came from Chicago to consult a little bit, to help us if we had problems. And we had a meeting with him and I had been doing some calculations and gotten some results. The calculations were so elaborate it was very difficult. Now usually I was the expert at this: I could always tell you what the answer was going to look like, or when I got it I could explain why. But this thing was so complicated I couldn't explain why it was like that. So I said to Fermi I was doing this problem and he said, 'Before you tell me the result, let me think. It's going to come out like this, because of so and so.' And it was a perfectly obvious explanation. So he was doing what I was supposed to be good at ten times better. He was quite a mathematician."

KLAUS FUCHS

"My wife died in Albuquerque. I borrowed Fuchs' car. He was a friend of mine in the dormitory. He had an automobile he was using to take the secrets away, you know, to go to Santa Fe. He was a spy.

I didn't know that. I borrowed his car to go to Albuquerque. The damn thing got three flat tires on the way."

In his book *Day One*, Peter Wyden tells how, "Women looked protectively upon Fuchs, the gentle bachelor . . . Children loved him and he was a favorite babysitter, usually available, ever reliable. Romantic attachments were so obviously lacking in his life, however, that the fun-loving Dick Feynman took him to task about his monastic existence as they sat on Fuch's G.I. bed in the austere Bachelor Dormitory No. 102, drinking orange juice. Fuchs frowned at Feynman's frivolousness and turned, as always, toward work. 'Don't you think the Russians should be told what we're doing?' he asked. Feynman nodded vaguely. It was not an unusual suggestion in those war days of close collaboration with the Soviets. 'Then why don't we send them information?' Fuchs insisted. Feynman said that such a decision was hardly up to them and he soon forgot what sounded like an abstract—but for Fuchs probably unique—outburst. By the time of the Trinity test, [The first test of the atomic bomb at Alamogordo, New Mexico, on July 15, 1945] Fuchs had fed the Russians atomic information seven times."[14]

VON NEUMANN

"The great mathematician gave me very good technical advice. We used to go for walks often to rest, like on Sundays, in the canyons of the neighborhood. And the one thing that von Neumann gave me was the idea that he had which is interesting, that you don't have to be responsible for the world that you're in. And so I have developed a very powerful sense of social irresponsibility as a result, and this made me a very happy man, since von Neumann put the seed in which grew now into my *ACTIVE IRRESPONSIBILITY*."

BOHR

"We were at a meeting and everybody wanted to see the great Niels Bohr. There were a lot of people, and I was back in the corner somewhere. Next time he was due to come I got a telephone call, 'Hello Feynman, this is Jim Baker (wartime cover name for Bohr's son). My father and I would like to speak to you.' 'Me? I'm Feynman.'

'That's right.' Okay. So at eight o'clock in the morning we go into an office in the technical area and start to talk. He (the son) says, 'We have been thinking of how to make the bomb more efficient and we think of the following ideas.' I said, 'No, it's not going to work.' He says, 'How about so and so.' I say, 'That sounds a little better but it's got this damn fool idea in it.' I was always dumb about one thing. I never knew who I was talking to. I would only worry about the physics. If the idea was lousy I said it was lousy. If it was good I said it looked good. A simple proposition. I've always lived that way. It's nice. It's pleasant if you can do it. I'm lucky. I'm lucky in my life I can do that."

"After this went on for about two hours of going back and forth over lots of ideas, arguing, the great Niels, always lighting his pipe, perpetually. It always went out. And he talked in a way that was un-understandable—'The questquestquest'—but his son I could understand. Finally, he said, lighting his pipe, 'I guess we can call in the big shots now.' So they then called in all the other guys and had a discussion. And his son told me what happened. Niels Bohr had told his son, 'Remember the name of that little feller in the back over there. He's the one guy that's not afraid of me and will say when I've got a crazy idea. So next time when we want to discuss ideas, we're not going to be able to do it with these guys who all say, Yes, yes, Dr. Bohr. Get that guy first. We'll talk to him first.' "

Though all scientists are curious, Feynman was more curious than most, rushing in where others feared to tiptoe. He volunteered to be hypnotized; took drugs, including LSD, to test their effect; and spent hours in an isolation tank to feel what it was like to hallucinate.

Some found him arrogant, which may account for the time he emerged from a fight in a seamy bar with a black eye. He was invariably arrogant towards interviewers, treating those he condescended to speak to as pests or idiots. When, for example, historian Robert Crease asked what seems a reasonable question: "Are scientists any closer to unification than in Einstein's days?" Feynman exploded in an angry harangue: "It's a crazy question. We're certainly closer. We know more. And if there's a finite amount to be known, we obviously must be closer to having the knowledge, okay? I don't know how to make this into a sensible question . . . It's all so stupid. All these interviews are always so damned useless." He walked out of his office shouting, "It's goddamned useless to talk about these things! It's a complete waste of time! The history of these things is nonsense!"[5]

When I interviewed him after phoning him at his home without warning, he was unpredictable as a subtomic particle, in turn amiable and irascible. Conversation with Feynman had an adrenalin arousing kick to it, not unlike combat.

I explained that as a non-scientist I was interviewing leading scientists hoping to get an insight into their minds and manners—in search of a mini-biography, so to speak. After a pause and a sigh, he replied, "It sounds like a lot of work, but go ahead." If I read him right, he was thinking there were worse ways of wasting his time though one didn't immediately occur to him.

THE INTERVIEW

I told Feynman: Robert Millikan said the more he studied, the more he was convinced of the existence of God. Linus Pauling said the more he studied, the less he was convinced of God's existence.

Feynman: I go with Linus. I'm talking about God as in organized religion. In that sense I'm convinced it's absolute nonsense. But if you make it abstract enough you can always make up something that doesn't mean anything—then it's all right. I usually put it this way: You define God and I won't believe it. Anything that has to do with ordinary religious ideas—no. For example, a man who came to earth and rises from the dead and so forth—absolutely not. I can only make a statement if I understand what the implications of the statement were, if you understand what I'm trying to say. If after making the statement you can draw some conclusions about the world from having made it, then I'll discuss it with you. But if you want to say: "Do you think the world is a gloobel?" and I don't know what a gloobel is, I can't say. And even if the word gloobel sounds like it means something, I'm not sure.

Let me put it another way: Are you in sympathy with people who say, "There seem to be definite rules for the world . . ."

Feynman: There are.

And we consider ourselves intelligent beings . . .

Feynman: [laughed] I guess that is also true, that "we consider ourselves . . ."

And some thing or force, at least equal to us in intelligence or creativity, caused our creation.

Feynman: Oh no, I don't see that at all. I don't know what that means:
"A force has to have the same intelligence." That stuff doesn't mean
anything to me. Now that doesn't mean I don't appreciate that the
world is very mysterious and wonderful, or that I understand it at
all deeply. There are many profound questions that leave me
awestruck and confused. All true. But that it has anything to do with
Protestant, Catholic, or Buddhist—no. Intelligence, no. Anything anthropomorphic, no."

Do you call yourself an agnostic or an atheist?

Feynman: An atheist. Agnostic for me would be trying to weasel out
and sound a little nicer than I am about this.

But I thought a scientist couldn't call himself an atheist, because that's
like saying "There is no God," and you can't prove a negative.

Feyman: I don't have to prove it. I only say: "Look, I don't know that
there is a God; I just don't think there is one."

That makes you an agnostic.

Feynman: No, no, no, no, no.

According to the dictionary (Webster's New World): an agnostic is
"a person who thinks it is impossible to know whether there is a God
or a future life, or anything beyond material phenomena."

Feynman: That's too refined. There's always an edge. What I mean
is this: the probability that the theory of God, the ordinary theory, is
right, to my mind is extremely low. That's all. That's the way I look
at it.

Has there been a great influence in your life?

Feynman: My father. Early in my life he'd tell me about the world,
about nature and how interesting it was. I don't know whether his
ultimate motive was just his interest in showing it to me, or whether

that was his way of convincing me to be a scientist. He didn't live to see me get the Nobel prize, but he was alive when I did the work that earned it.

Is your essential interest the urge to uncover secrets?

Feynman: That's what I usually say. I don't know enough psychology to know my true motivation, but that's what I think it is. But I might be happily fooling myself.

Is that why you were hypnotized?

Feynman: Yes, because I wanted to find out what it's like. I'm a fall guy for such experiences. I was a graduate student at Princeton and they announced that a lecturer was coming to lecture on hypnosis and needed some volunteers. The dean was telling us about it at dinner in the great hall, where we were sitting around in academic gowns. He was over at one end of the dining table in the big room and I was way down the other end. And I thought, ah, dammit! He kept talking about how he was going to ask for volunteers, and (I thought) he's not going to hear me back here where I'm sitting. And I wanted very much to volunteer because I was very interested to learn from such experiences. So, finally, he gets around to saying, "Anyone want to volunteer?" So I shot up my hand, flying out in the air, back down there at the other end of the hall and then, "MEEE!" I said. There wasn't another soul jumping up. I was all by myself down there. He says, "Yes, Feynman, we know that you'll probably want to do it. Now, who else?" Anyway, that was the first time I got hypnotized. After I found I could be hypnotized he used me for a demonstration lecture there. And he made a post-hypnotic suggestion that when I went back to my chair I would go around the audience instead of walking in a direct line. I started walking directly, because I knew I had this urge in me and was going to defy him. But I felt so terribly uncomfortable that I had to turn around and walk all the way around the audience. A very interesting phenomenon.

It could be used to improve your memory and tap your unconscious or subconscious.

Feynman: Your view of the world is entirely different from mine. I'm not trying to find a way to improve myself. I do all these things for curiosity about the phenomenon. I'm trying to find out more about hypnosis.

But maybe there's information in your unconscious that hypnosis could untap and help you win another Nobel prize.

Feynman: Oh, sure, that's perfectly possible. Anything can happen in my unconscious mind. [His tone implied the unconscious mind was a minefield with possible treasures, but he wasn't keen to reconnoiter.]

But wouldn't you want to reach it through hypnosis?

Feynman: I don't care. Perhaps you can tell me how to do it.

I'll try to find you a good hypnotist. [His reluctance still puzzles me. What had he got to lose? After all, he had tried drugs in a vain attempt to stimulate his thinking.] Is it true that you broke the Los Alamos security code and opened a safe containing top-secret documents? Then left behind a note that said "Guess Who?"

Feynman: When I was at Los Alamos one of my hobbies was to try to open safes and locks, a sort of locksmith-type hobby. Practicing opening locks, I at one point opened the lock of the safe that contained all the secrets of the atomic bomb, and the whole business behind them. There were nine filing cabinets containing all the documents at Los Alamos. I opened three of them to check if they all had the same combination. I left notes in them to tell the guy that he shouldn't have locks with all the combinations the same, and stuff like that. And that I'd taken the documents out. And there were certain jokes in my notes. I was standing in the office there playing with the safes in full light of day. The guy who was running the office was a friend of mine. And he was very upset when he found the safes had been opened. They probably changed the combinations after that.

[A nice ironic touch: Here was Feynman playing about with secret documents. One night he got drunk and a fellow physicist drove him back home. The driver was Klaus Fuchs. Feynman probably knew Fuchs as well as anyone at Los Alamos. They once joked about which of the two of them would be the most likely candidate as a suspect for possible espionage, as Feynman later told an FBI agent. They agreed it would be Feynman. Fuchs was later revealed to be a spy for the Soviets who had passed atomic bomb secrets. He confessed and went to prison. None of the secrets leaked through Feynman. He handed back all the documents he'd taken out just for fun.]

Do you think Robert Oppenheimer got a raw deal when he lost his security clearance?

Feynman: Yes. Not just losing his security clearance. That was just a symbol of it. Yes, definitely.

But he was mixed up with communists and with Chevalier.*

Feynman: That may be true. But at the time people [among them General Leslie Groves, head of the atom bomb project] knew about that, and had reviewed it when he was at Los Alamos. It was clear to me at the time they brought it all up it was a political problem. [I should have asked Feynman to explain what he meant. I assumed he meant internal politics, especially the incompatibility of Oppenheimer and his nemesis, Edward Teller. Feynman himself had aroused suspicions at Los Alamos through his wacky behavior. I questioned him about it.]

How did you confuse the censor during World War II?

Feynman: I had a wife that was [fatally] ill in Albuquerque, in bed with tuberculosis. She didn't have much to do. And she was writing me letters, and saw an advertisement for a jigsaw-puzzle blank. You put it together and write the letters—and then you break it up and put it in a little sack, and send it. And she had done that, that's all. [To amuse him and entertain herself. The fact that it also aroused the curiosity of the censor added to the fun for Feynman. Knowing of his

---

* Oppenheimer told General Groves that Haakon Chevalier, a language professor at Berkeley, had discussed with him the idea of sharing atomic bomb information with their allies, the Russians. Chevalier denied it. Testifying before the House Un-American Activities Committee in 1949, Oppenheimer said of their conversation, "Dr. Chevalier was clearly embarrassed and confused, and I, in violent terms, told him not to be confused and to have no connection with it. He did not ask me for information." Oppenheimer's former girlfriend, Jean Tatlock, was a Communist Party member. So was his younger brother, Frank. In the 1930s, Oppenheimer himself had contributed to organizations connected with the party. Between 1942 and 1955 counterintelligence agents followed Oppenheimer, tapped his phone, read his mail and bugged his office and homes. No evidence was ever presented questioning his loyalty to the U.S.

wide interests I thought they might extend to cosmology, which explains my next question.]

Have you any interest in the possibility of intelligent life on other planets?

Feynman: Oh, yes. But I think we know nothing about it. We don't even know how to estimate the odds. There are, of course, the usual things to say: That there are many positions in space and time where such things might have been similar to earth. I'd like to find out, but I don't have any answers or even any odds.

Any views on UFOs?

Feynman: Just one of those crazy things. False belief. It comes from uncareful observation.

And the accounts people give of being abducted by aliens are the result of hysteria?

Feynman: Even worse. I don't know what you'd call it. I'm not a psychologist, but it has nothing to do with reality. Though it might be *their* reality. In my opinion, no one has been taken aboard a spaceship. I think it's extremely unlikely. In fact, that's the way I look at the whole world. I have no certainty. It's just varying degrees of probability. But when the probability gets sufficiently small I consider it for practical purposes as being untrue. Therefore, rather than speak of myself as an agnostic, which gives the impression I'm still debating the question, I'd rather say the odds are way over on the one side there, so high . . . Like I'm here in California and you're in Florida? I don't know where the hell you are. So the odds are not very high. You might be right here across the street for all I know. Go ahead.

How do you relax?

Feynman: I jog about six miles in the morning. I used to draw. I've done biology experiments and I've deciphered Mayan hieroglyphics. One of my hobbies is curiosity about how dream images and hallucination images are formed. And how the brain works. I had not much opportunity to see hallucinations; it's a very rare thing. When I heard you can have hallucinations in an isolation tank I thought it was a great chance. So when I met the man, John Lilly, who told me he was using isolation tanks I said, "Can you get hallucinations?" He said, "Oh, yes." And I was so excited. He let me spend two or three hours at a time in his tanks. And I had about a total of twenty-four hours trying to get hallucinations. I did, ultimately, after two or three two-hour sessions. I learned what hallucinations meant and a little bit how they worked. It was very interesting. It's something like dreams. One hallucination involves the idea that you leave your body—that is, the location of "I"—which you really think of as up near the top of your head, behind your eyes. But that location is a matter of convention. As a matter of fact, the Greeks didn't even know where the seat of reason was, and argued about it. So it may not be that you have to have it behind your eyes. One of the hallucinations is that "I" is located somewhere else, and you can even move it out of the body. You have other hallucinations where you fly over scenes and see things. First I moved down and out through the hips somewhere. But other times I had different hallucinations when I'd fly above scenes. Every once in a while the system collapses and I'd fall down and see the scene from normal level, and try to get up there again. It wasn't at all worrying or frightening. The opposite. Because I wanted to see hallucinations I was happy to see them. In those days I was giving lectures in a class near the place where John Lilly had the isolation tanks, so it was very convenient for me to do it once a week.

Was the Nobel prize a big surprise?

Feynman: Yes. Because, like any human being, when Nobel prize time came around I used to fantasize and daydream a bit. But nothing would happen. Then I lost the habit. When it happened it was a complete surprise. I didn't even know it was the right season.

Did you do anything unusual with the Nobel prize money?

No. [He did buy a beach house in Mexico after winning the award.]

How many hours a day do you work?

Six or seven. Sometimes if the ideas are good, over the weekend I'll take the work home. Or when we go to the beach to relax, I'll take some stuff with me and I'll think hard about some problem. Almost all of this thought ends up in the wastebasket. But I do an awful lot of thinking. I guess my wife and children ought to tell you how much

I actually work. It's hard to tell. You're driving and you're thinking, and then you're not thinking sometimes.

Any major disappointments?

Feynman: Oh, indeed. I've spent years trying to do things I didn't succeed in doing. I've worked on the theory of turbulence without success.* I guess one place where I felt a little bit uncomfortable was after I worked on liquid helium, I worked on superconductivity for a couple of years and was unable to understand it.

Something you should have grasped?

Feynman: Yes.

And that a lot of other scientists did grasp?

Feynman: Not a lot, no sir. It's been worked out by Bardeen and Cooper. It wasn't easy. Somebody did get it, but not me. On the whole there were not many of those. It wasn't a very severe disappointment, just something I remembered when you asked for it. I got mostly, almost always, pleasure. As a matter of fact, even working on superconductivity was a lot of fun, and I worked very hard and had a good time with it. I enjoy working even if I don't solve anything.

When do you know you're not going to solve a problem and decide to quit?

Feynman: The biggest trouble is the gradual growth of confusion. It's hard and you get mixed up and confused. It gets kind of sticky. New ideas come more and more slowly, and it gets dull. You struggle on in this dull part for a while, this quagmire. Maybe you think of another problem and get interested in it, and leave the first one alone. It's rather hard to leave something alone, but that's the way it ends up.

What's been your most interesting discovery or work since the Nobel prize?

---

* Turbulence, an aspect of chaos, is the last unsolved problem of classical physics. One example is the smoke rising vertically from a cigarette that suddenly disrupts into irregular whorls.

Feynman: My prize was granted for something in 1947, although it was given in 1961, I guess it was. [He received it in 1965.] Next, I had two pieces of work, either one of which I liked very much. One was the theory of liquid helium. [He came up with conceptual wave functions that describe liquid helium.] I had a great deal of pleasure working that out. And another one was to discover the laws of weak interaction in about 1954 with Mr. [Murray] Gell-Mann. That was also exciting. [The Feynman–Gell-Mann team predicted new kinds of radioactive particle decay, later experimentally confirmed.] Since the actual award of the Nobel prize I guess I worked out something called the theory of partons to explain some of the properties of protons. And that worked out very well. [His fresh interpretations of electron-scattering experiments suggesting that free floating subunits, his "partons" were deep within proton and neutron. It opened one main experimental path towards quarks—from which everything in the world is made.] Right now I guess I'm working on nothing. I'm standing back. I have a leave coming up, and I just want to let myself play around with ideas and I don't know where the hell I'll come out.

Soon after I spoke with Feynman he was found to have abdominal cancer. He underwent four major operations, as well as radiation, hyperthermia, and other treatments, which prolonged his life for several years. He was still teaching at Caltech two weeks before he died on February 15, 1988, at the age of 70.

Morphine didn't dull his spirit, and he even managed a quip at the end. His wife, Gweneth, was at his bedside, and his last words to her were, "I'd hate to die twice, it's so boring." She died of cancer less than two years later, in December 1989, at the age of 56.

Helen Tuck was Feynman's secretary at Caltech for sixteen years. She told me: "He spent a lot of time in my office. And Murray Gell-Mann's office is on the other side of me. I always felt like the thorn between two roses. Sometimes the arguments would go on in my office so it was a very interesting time. They were both really great. They would have lively discussions and sort of poke fun at each other. Murray Gell-Mann has a fetish about pronouncing every word 100 percent correctly and names, even foreign names he always has to put his own twist on them. And Feynman didn't give a damn about that. He believed words were to be understood, not rolled around your tongue in some language so you couldn't understand it when it came out. One time Murray was saying something about Montreal and he gave it a very French twist and Feynman said, 'I thought words were to be understood.' Because Feynman didn't have a good ear for things like that."

I picture him as a spellbinding lecturer. But didn't a lot of students walk out on him?

Tuck: He would become so enthused that you could become a little glazed and maybe some students were unable to cope with it. He soon toned that down and Physics One, a course he made popular years ago, became a very popular course on campus.

He had no time for philosophy, I understand.

Tuck: One time the philosophers' society was having a big meeting in San Francisco and he finally accepted an invitation to be a speaker, with the very clear knowledge he was going to give them his very strong negative opinion. Because he didn't believe in what they did. [Philosophers were not rigorous enough for his taste.] And he pretty well emptied a room at the meeting.

Although he loved his father, when he was buried Feynman refused to say a prayer over the grave. Walked out in fact when the rabbi pressed him. Can you explain that?

Tuck: He once told me he was the only anti-Semitic Jew I would ever meet. His sister's faith is pretty traditional. We have a lot of visitors in our department from Israel and he didn't believe in a lot of the things they were doing.

Was he against the state of Israel?

Tuck: I wouldn't 100 percent say that. But he wouldn't go there and he had many offers to go with his whole family. And I'd say, "Gosh, I'd give the world to be able to go."

He went to Tibet.

Tuck: Well he never would go to Israel. It was really strange.

You don't mean Feynman was anti-Semitic in the way the Nazis were?

Tuck: Oh no, nothing like that. He believed in live and let live.

It's just that he wasn't religious, and the country of Israel didn't appeal to him?

Tuck: That's right. And I don't think he followed their politics.

Because he instead, devoted himself to science and hobbies like translating Mayan hieroglyphics?

Tuck: And his family.

[To counter Feynman's claim to be anti-Semitic are his words: "Another guy wanted me to join some kind of club of professors. The club was some sort of anti-Semitic club that thought the Nazis weren't too bad. He tried to explain to me how there were too many Jews doing this and that—some crazy thing. So I waited until he got all finished and said to him, 'You know, you made a big mistake. I was brought up in a Jewish family.' He went out, and that was the beginning of my loss of respect for some of the professors in the humanities, and other areas, at Cornell."[16]]

Did you detect any mistakes in *Genius,* Gleick's biography of Feynman?

Tuck: In my office opposite my desk is an alcove where I have two guest chairs and Feynman would sit there for hours at a time with his feet propped against my desk. And he'd do his work and I'd do mine. And in Gleick's book all he says is that "his secretary, Helen Tuck, says he used to hide behind the door in her office." And I thought it made him look so foolish. He didn't hide behind my door.

Knowing he tried to avoid interviewers and shouted at students, "Go away, I'm busy!" it gives a picture of the man wanting to concentrate on his own work and not being the kind of professor who invites students to his home. And because he wanted to get on with his own work, hiding behind your door I understood to show he neither wanted to hurt the person he was avoiding, nor waste time.

Tuck: He was always open to students. Once in a while, if someone rapped on his door while he was preparing for class, he would yell, "Go away, I'm busy!" And my office being right next door and my door being open, I'd say, "He's getting ready for class." That's all there was to it. After class three or four kids would come back with him and they'd be in the office for two hours. He used to get a kick out of it when his phone rang and I'd say, "I'm sorry, Dr. Feynman isn't in his office" and he thought it was funny—because he was sitting in my office. And people would stand in the doorway and ask if he was in his office and I'd say, "No," and they'd go away. They could see a man's feet and legs were here in my office but didn't know it was Dr. Feynman. I got to be pretty knowledgable about who he'd see and would just tell them to come on in. But if it was somebody strange, like a publisher, I'd let them wander off and not bother him.

I presume he didn't have a religious funeral.

Tuck: It was a very simple one. It was not a Jewish service. But the year after he was buried his sister called me to ask for help in lining up ten men of the Jewish faith to be at a service over his grave.

Speaking with Kip Thorne, a Caltech physicist, I said: In *Genius,* James Gleick's biography of Feynman, he says Feynman was so tough on you that you broke down in tears.

Thorne: That was quite off the mark. Not accurate. The story on myself was that I gave a physics colloquium and Feynman was a close personal friend of mine and pressed me hard about some issues related to the colloquium that I hadn't thought about for a year and a half. And Feynman was a lot quicker mentally than I was and I wasn't able to give an adequate response. I was pretty upset with myself and I was in fact sick that night over being so upset with myself that I hadn't given the proper response. But the twist placed on it by James Gleick in his biography is one that illustrates a phase of Feynman's life in terms of his relationships with his colleagues. I think that's not correct. I think this was an isolated incident and not an illustration of anything of the sort that Gleick claims.

Wasn't Feynman tough on anyone who didn't come up to his intellectual level?

Thorne: He was tough on people if he thought they could stand it, which he thought I could. We were close friends with a lot of respect for each other. Feynman would become excited by ideas and press hard with regard to them, paying little attention to what effect his pushing was having on the person he was talking to. If he later became aware he had caused pain he got quite upset. He would become rather oblivious to that issue often in the excitement of the intellectual give and take. He is portrayed in this sense as being similar to what Lev Landau was. And Landau, I'm sure, was truly that way. I've a number of close mutual personal friends of Landau's. But that was not the nature of Feynman's character.

How do you class Feynman among his contemporaries, Murray Gell-Mann, for example?

Thorne: They were very different kinds of scientists. The issues are not one dimensional. There are senses in which Feynman was head-and-shoulders over Gell-Man, and there were other senses when he was not as good as him.

How would you compare John Wheeler and Feynman?

Thorne: Wheeler was Feynman's chief mentor [at Princeton]. They both had a tremendous impact on physics.

Can you compare Feynman with Einstein?

Thorne: Again, these things are very multidimensional. Certainly Einstein had the biggest impact on physics of anybody in the 20th century. Feynman, in that respect, is not as great. One of Einstein's greatest strengths was his physical intuition. Feynman was probably the only physicist in this century whose physical intuition was comparable to Einstein's.[17]

Chapter Three.   Paul Dirac

Brian: Dirac, after Einstein died, was called the greatest living physicist.
Linus Pauling: I don't like to make statements of that sort,
but I can accept that all right.

Paul Dirac had a peculiar, constricted upbringing, living in Bristol, England, with his Swiss emigrant parents. The Diracs never had guests in their home and their three children—Paul, an elder brother, and younger sister—were discouraged from making friends. Outside the house, said Dirac, "I didn't speak to anybody unless spoken to. I was very much an introvert, and spent my time thinking about problems in nature." He didn't speak much inside the house, either.

His domineering father was largely to blame. He taught French, and urged Dirac to become bilingual by addressing him exclusively in that language. But "since I found that I couldn't express myself in French, it was better for me to stay silent than to talk in English. So I became silent at that time—that started early." And he remained, lifelong, a man of few words and long pauses.

The family ate their meals separately: Dirac's mother in the kitchen with her eldest son and daughter, Dirac with his father in the dining room. Dirac's odd explanation for why he didn't eat in the kitchen, which he would have preferred, was that there weren't enough chairs there.

Instead of a linguist, Dirac's father had produced a son tongue-tied in two languages. He changed his tactics then and prodded the teenager toward the solitary science of mathematics.

Contemplating careers as engineers, both Dirac and his elder brother took on-the-job training at a factory, maintaining the family atmosphere by not exchanging a word. The brother later committed suicide.

Dirac's flair for math got him into the honors program at Bristol University, where his reluctance to talk steered him from engineering to physics. The only other student in the program was a woman who insisted on studying math which could apply to physics. Rather than discuss it, Dirac silently followed her lead.

From Bristol he went east to Cambridge University where, in 1926, he received a Ph.D. for a thesis on quantum mechanics. Impressed with Dirac's work, the father of quantum mechanics, Niels Bohr, invited him to Copenhagen for further study. There Dirac's colleagues, Werner Heisenberg, Wolfgang Pauli, and George Gamow, also destined for greatness, often engaged in furious arguments about life, literature, and the pursuit of the atom.

Dirac was conspicuous among this exuberant group by behaving as though he had taken a vow of silence. "I admired Bohr very much," he said later. "We had long talks together, very long talks, in which Bohr did practically all the talking."[1] Since Dirac rarely initiated a conversation and usually answered question in monosyllables, when Bohr was asked how the Englishman was doing he couldn't say. Overhearing Bohr's comment, J.J. Thomson, the electron's genial discoverer, and Master of Trinity College, Cambridge, said: "That reminds me of the parrot that wouldn't talk. When his owner went back to the pet store to complain, the storekeeper explained: 'So you wanted a talker: I sent you a thinker.' "[2]

Einstein seconded that assessment when he met Dirac in 1927, at the Solvey Conference in Brussels, Belgium. At 25 Dirac had established himself as one of the world's most important physicists based on his seminal paper, "The Quantum Theory of the Emission and Absorption of Radiation." In Einstein's opinion it gave the most logically perfect presentation of quantum physics."[3] Unexpected praise from one who resisted quantum physics. And a dramatic change from the year before when Einstein had said Dirac was "balancing on the dizzying path between genius and madness."[4]

Occasionally Dirac surprised fellow physicists by joining in their lively discussions. Once he fervently attacked a society in which the poor suffered, and the greedy, along with organized religion—which he called a ludicruous sham—prospered. But usually he kept his thoughts to himself and was regarded as rude, which he wasn't, or reserved and antisocial, which he was.

In 1928, walking alone through fields near Cambridge he might have been mistaken for a poet retracing the footsteps of Rupert Brooke. He *was* a poet in a way, seeing poetry in mathematics and beauty in equations that might explain the universe. Though Dirac expressed his attitude less fancifully: "One must follow up a mathematical idea and see what its consequences are, even though one gets led to a domain which is completely foreign to what one started with."

On one of his frequent solitary Sunday walks he sought to reconcile Einstein's relativity and Max Planck's quantum theory, guided by the notion that the beauty and symmetry he found in mathematics would be reflected in nature. He returned from the walk startled by his idea that to be symmetrical the universe must contain not only matter but its opposite, antimatter. Dirac feared his thinking must be flawed and that he had wandered into a domain more appropriate for science fiction or fantasy. Nevertheless, despite qualms, he published his theory.

His scientific peers reacted with indifference or alarm. Among the latter was Heisenberg who wrote to Pauli: "The saddest chapter of modern physics is and remains Dirac's theory." Heisenberg explained, that until 1928 he had the impression that "in quantum theory we had come back into the harbor, into the port. Dirac's paper threw us all into the sea again."[5] Dirac was among those looking for a life raft, well aware of the gaping hole in his theory: it required the acceptance of antimatter, which had never been shown to exist.

Meanwhile, help was at hand. At Caltech, Carl Anderson, guided by Robert Millikan, had built a cloud chamber with the world's highest existing magnetic field. A cloud chamber supersaturated with water vapor revealed the presence of moving charged particles. Anderson intended to use it to study cosmic rays. Instead, in 1932, he accidentally discovered antimatter or the positron, as it was called, in his cloud chamber, vindicating Dirac's theory. He also observed the devastating effect of contact between matter and antimatter that destroy each other in a flash, releasing energy in the form of a photon. Heisenberg later acknowledged that the discovery of antimatter was perhaps the 20th century's greatest advance in physics.

The same year Anderson discovered antimatter, Dirac was appointed to the chair of mathematics at Cambridge, a chair once held by Isaac Newton and presently held by Stephen Hawking.

Dirac's equation that embraced antimatter explains the mechanics of the atom, and earned him the 1933 Nobel prize in physics, which he shared with Erwin Schrodinger.

George Gamow, memorable for the "Big Bang" or expanding universe theory, saw Dirac as a nutty professor whose discoveries took science into a ribald, Alice-in-Wonderland world in which "an electron interferes with itself." (Though Gamow himself was known among colleagues as "goofy Gamow.") In fact, Dirac's thinking led to television, computers, and space travel.

Fame and success had little apparent effect on Dirac. He still treated what he considered excess in speech or writing with dry disdain, and preferred to work alone because, he said, then he could keep his mistakes to himself. When Dirac was briefly visiting Berkeley, Robert Oppenheimer persuaded him to listen to the theories of the young physicists, Robert Serber and Arnold Nordsieck. Dirac listened for an hour without asking question or making any comment, except to ask them, at the end, the way to the post office. Somewhat desperate, Serber asked if he might accompany him to hear what he thought of their ideas. He turned them down, saying he couldn't do two things at once! This was an echo of his purported criticism of Oppenheimer's interest in poetry, "How can you do both—poetry and physics? In physics we try to tell people things in such a way that they understand something that nobody knew before. In the case of poetry it's the exact opposite."[6]

One of Dirac's achievements was the path-integral formulation in quantum physics, which Richard Feynman developed in the 1950s. Both men first met at the 1961 Solvay Conference, where the brash, buoyant, and usually irrepressible Feynman hoped to break down the taciturn Dirac. Abraham Pais happened to witness the encounter and recorded their exchange as follows:

Feynman: It must be wonderful to be the discoverer of that equation.

Dirac: That was a long time ago. [lengthy pause] What are you working on now?

Feynman: Mesons.

Dirac: Are you trying to discover an equation for them?

Feynman: It's very hard.

Dirac: One must try. [end of conversation]

After retiring from Cambridge in 1969, Dirac settled at Florida State University in Tallahassee. An odd choice when the world was open to him. But he explained its appeal to me: it had the best physics department in the southeast, a daughter lived nearby, and he wouldn't have to drive on icy roads.

At Tallahassee he explored his idea that gravitation weakened as the earth ages. Asked why it might weaken, he replied curtly, "Because God made it so!" Then, fearing he might be misunderstood he quickly added that he did not subscribe to any organized religion.

Paul Dirac was ranked by *Compton's Encyclopedia* in 1975 as probably the world's greatest theoretical scientist, on a par with Newton and Einstein. *Science News* did not dispute this, but called him "a cranky, downright ornery man." Of course the two views are quite compatible, even have a sort of symmetry.

Dirac carried the scientific method into his reading of fiction. He was curious to know what had really happened in the cave in E.M. Forster's *A Passage to India*. When asked about it, Forster, who was at Cambridge with Dirac, gave different answers at different times. Finally, Dirac asked: "Could there have been a third person in the cave?"

"As usual," said Rudolf Peierls of Oxford University, "Dirac had thought of an interesting possibility which had occurred to nobody else."

I heard that Dirac brought the same mindset to his reading of Dostoevski, and I asked him about it, catching him in a rare, expansive mood.

After lending you Dostoevski's *Crime and Punishment*, your friend Peter Kapitsa asked how you liked it. You apparently replied: "It's nice, but in one of the chapters the author makes a mistake. He describes the sun as rising twice on the same day." Is it true that this was your only comment on the novel?

Dirac: It was setting and not rising. It occurs rather at the beginning of the book. He describes a sunset, and then a little later the same evening the sun sets again. That kind of mistake does jar on me.

I understand. The intent of the anecdote though, is to stress that after reading a great work of literature you had no comment on the artistry, the characters, or its emotional effect. Is that true?

Dirac: I don't remember.

Did you know the physicist Wolfgang Pauli?

Dirac: Fairly well, yes. There's a well-known joke about his riding in a train which stopped at a station for a few minutes. At the same time in a nearby laboratory, apparatus for the study of atomic phenomena collapsed. It was suggested that Pauli caused the collapse by a poltergeist or psychokinetic means. Gamow called it "The Pauli Effect." ["The Pauli Effect: A mysterious phenomenon which is not and probably never will be understood on a purely materialistic basis." George Gamow.] Oh yes, Pauli was supposed to bring bad luck. [He chuckled.]

I take it you are a skeptic.

Dirac: Yes. If there was such a thing as telepathy, for example, it would have been used by spies in wartime. They never have used them.

How do you know that?

Dirac: I've never heard of them using it. Telepathy would be the ideal method for trafficking in information. When I was a research student at Cambridge in the 1920s, a young man there named Gatty was interested in telepathy. We did experiments together to see whether we could get any results. We sat in a room, and one person would think of a number, one to ten, and write it down. The other person would try to think what the number is. Sometimes we got a long series of successes. Later, we changed it to drawing a number from a pack of cards. Then there was no agreement at all. Our explanation for the successes with our first experiment was that after seeing a certain sequence of numbers another number may suggest itself, and the other number may be the same for both of us. Just recently I heard an explanation of how Beethoven composed his work. He got a sequence and then found out what sequence of notes had to follow the previous sequence. I think our sequence of numbers was like that. Seeing a sequence of numbers suggests another sequence of numbers to come after. [That seems more improbable than telepathy!] Anyway, we concluded after a few months that there was no telepathy effect.

How about the hundreds of reports of people worried that someone close to them is in trouble or danger at a specific time, and it turns out to be true? Or accounts of dream premonitions that are accurate? Do you dismiss these as coincidences, hallucinations, or what?

Dirac: There may be some reason behind it. You have a good friend who drinks too much. You might suddenly get the fear that he'll have some trouble from his drinking, and that trouble may really occur. When we did those telepathy experiments we tried not to have any preconceived ideas. I was open-minded. But the results were negative, and the experience convinced me there was no telepathic effect. My conclusion was that one could easily be misled. If a distant clock strikes a certain number, that number may suggest itself to both of us and we write it down. That's the sort of error one may get into. One can easily be mistaken.

You've said that all simple problems have been solved. What difficult ones still baffle you?

Dirac: I am still trying to find the right, beautiful equations to describe the atom.

Why beautiful?

Dirac: Because a good equation must be beautiful. Some people can appreciate beauty, others cannot; it's like appreciating anything beautiful.

How would you describe your scientific achievements?

Dirac: As well as my work on the quantum theory, I contributed to research on electromagnetic fields and in elementary particles. [Banesh Hoffmann and Helen Dukas express his achievements more colorfully: "Dirac, in a particularly felicitous way, had shown that the new quantum mechanics was essentially Newtonian mechanics with a quantum transfusion. In 1928, Dirac brilliantly applied the special theory of relativity to the quantum theory of the electron, an achievement as remarkable for its mathematical beauty as for its spectacular success. For these and other achievements he was awarded the Nobel prize."][7]

What interests you now?

Dirac: I've been working on the idea of gravitation, whether the gravitation of a constant is varying. [He theorized that as the earth aged, gravity weakened.] I've done a great deal of work on that. It's not altogether proven, but it is of interest. Many people are working on it.

A complete explanation of the universe is how far off?

Dirac: It may take hundreds of years to find the answer. Maybe there is no final answer.

But, of course, you scientists will keep trying.

Dirac: Yes.

[I knew Dirac had changed his attitude to religion from ridicule to respect. As Heisenberg recalled, the 25-year-old Dirac had once broken his usual silence to say: "Religion is a jumble of false assertions with no basis in reality. The very idea of God is a product of human imagination. I can't for the life of me see how the postulate of an almighty God helps us in any way. What I do see is that this assumption leads to such unproductive questions as why God allows so much misery and injustice, the exploitation of the poor by the rich and all the other horrors He might have prevented."][8] In 1983 at age 80, Dirac told an Associated Press reporter he agreed with Pope John Paul II that science and religion were "both seekers after the truth. There cannot be any conflict between people who are seeking the truth even if they seek it by different methods." When I called back to ask him if he had changed his mind radically about God and religion, his widow answered the phone. He had died a few months previously, on October 20, 1984.]

MRS. PAUL DIRAC

What was your husband's attitude to religion?

Mrs. Dirac: He was a member of the Pontifical Academy at the Vatican It is an international academy of seventy invited member regardless of creed, among them the most distinguished scientist in the world. Its goal is to foster pure science.

I gathered that Dr. Dirac's attitude towards religion and God was similar to Einstein's: he had a religious sense but wasn't a believer; wasn't a Christian, for example.

Mrs. Dirac: He was a Christian. He went to church on Sundays.

You mean he believed in Jesus Christ?

Mrs. Dirac: Perhaps sometimes, and sometimes not. You know, most people are like that.

Most scientists I have contacted are atheists.

Mrs. Dirac: My husband wasn't an atheist. In Italy, once, he said, "If there is a God, he's a great mathematician."

Ah, if there is a God. He did say if. I understood Dr. Dirac might have been buried in Westminster Abbey. [along with Isaac Newton]

Mrs. Dirac: Yes, but that would have been a great palaver and he was a very simple man, very modest and very unworldly. Who else are you interviewing?

Richard Feynman, Linus Pauling . . .

Mrs. Dirac: Linus Pauling adores the limelight. [She chuckled.] I'm sure he loves to be interviewed.

Both Pauling and Feynman said they didn't believe in God.

Mrs. Dirac: My husband never said that.

Did he feel there was an intelligent creator?

Mrs. Dirac: Yes, yes.

And anticipate life after death?

Mrs. Dirac: I don't think he thought much about it.

Did he believe there might be life on other planets?

Mrs. Dirac: He was very much for space research.

He told me that in his early years at Cambridge University he conducted telepathy experiments and concluded telepathy doesn't exist.

Mrs. Dirac: I never heard that. But I have experienced telepathy. For me it happens, and I believe very much so.

You have had experiences that convinced you ESP exists?

Mrs. Dirac: It does exist. I've had some experiences: thinking of my daughter and she telephones. But even more so, forebodings that something will go wrong, and it does go wrong: an illness or a very bad, difficult time. I discussed this with my husband, but he didn't make any remarks about it. He wasn't a great talker. [Extraordinary, though, that he had never discussed his telepathy experiments with her.]

What did you think of George Gamow's anecdotes about your husband in *Thirty Years That Shook Physics*?

Mrs. Dirac: I don't read science fiction.

According to Gamow he reported the following anecdote "for the benefit of future historians." And it was this: At a party in Copenhagen, Dr. Dirac proposed a theory that there must be a certain distance at which a woman's face looks its most attractive. But when questioned by Gamow, Dirac admitted the closest he had ever been to a woman's face was about two feet.[9]

Mrs. Dirac: That's very unlike him.

Also according to Gamow, while watching Anya, Dr. Kapitsa's wife, knitting a sweater, Dr. Dirac got interested in the topological aspect of knitting, and said that he had discovered from observing her that there were two ways to knit. One was the way Anya was doing it. Then he began to demonstrate the other way, until she told him what he was doing was well-known as "purling."[10]

Mrs. Dirac: I don't think that's true.

Here's more from Gamow. Dr. Dirac had a reputation as a man of few words who hated redundancies of any kind. He was once handed a manuscript on which was written: "Must not be published in any form." And Dr. Dirac erased the words "in any form." Does that sound like him?

Mrs. Dirac: I guess so.

Finally at the end of one of Dr. Dirac's lectures he invited questions, and a student said: "Dr. Dirac, I don't understand the formula on the top left side of the blackboard." Gamow said that Dr. Dirac replied, "That's not a question. It's a statement. Next question, please."[11]

Mrs. Dirac: That doesn't sound true.

How do you explain these anecdotes?

Mrs. Dirac: Gamow was a very childish man. He was a great friend and admirer of my husband, but I think he invented some of these things.

Everyone agrees your husband was a man of few words.

Mrs. Dirac: Absolutely true.

Although when I spoke with him we had a good conversation.

Mrs. Dirac: [She chuckled.] He mellowed with age. He gave very few interviews. He was afraid of publicity and avoided it as much as possible. He was a very private person and didn't like the limelight.

Did your daughters ever get their father to talk?

Mrs. Dirac: Yes. If it was science, he was very easy to talk to. But I don't think they ever tried to talk to him about personal things, because I was here and they could do it with me.

He was the quintessential shy, introspective man.

Mrs. Dirac: Yes, very shy, very private, and very withdrawn.

Was he happy?

Mrs. Dirac: When his work went well, yes he was happy.

What was his routine?

Mrs. Dirac: He worked all the time. That was his hobby. When he stayed in Cambridge, England, he only went out to give lectures. Here in Florida he went to the State University at Tallahassee like a bank director. From morning to evening he was at the university and he studied and worked there.

Did he regard himself as British?

Mrs. Dirac: Yes. He was a British citizen and he had Swiss citizenship from his father, but he never used it.

Who were his heroes in the scientific world?

Mrs. Dirac: Peter Kapitsa, the Russian scientist. [Famed for his investigation of matter at extremely low temperatures and under the influence of superpowerful magnetic fields. He declined to take part in either the Soviet's atom bomb or hydrogen bomb project and lived to tell the tale.]

What did Dr. Dirac think of Einstein?

Mrs. Dirac: He loved Einstein and he had a very strong admiration for his work and character. He loved Peter Kapitsa, too. He had known him well at Cambridge.

Dirac kept a photograph on the wall of his office in the physics building at Florida State showing the scientists who attended the Solvay Physics Conference in Brussels in 1927. Among them were Marie Curie, Erwin Schrodinger, Niels Bohr, Werner Heisenberg, Albert Einstein, and Paul Dirac. Dirac, sitting almost directly behind Einstein in the photograph was, when I spoke with him, the sole survivor of that group who made their momentous discoveries during the Golden Age of theoretical physics.

Among Dirac's last words was a testament to Einstein: "He did wonderful work all alone, not needing much money for it, only enough to live." The same could be said of Dirac.

Chapter Four:
Victor Weisskopf

"I often tell my students when they are depressed by the world,
that there are two things that make my life worth living:
Mozart and quantum mechanics."
—*Victor Weisskopf*

Victor Weisskopf worked with Niels Bohr and Wolfgang Pauli, helped build the atom bomb, made music with Albert Einstein, discussed communism with Robert Oppenheimer, crossed swords with Edward Teller, and taught physics to Nobel laureate Murray Gell-Mann. His autobiography,[1] wrote *Daedalus* editor Stephen Graubard, "suggests that a great research physicist can also be a great humanist." The *Boston Globe's* Russell McCormack called him, "one of the most thoughtful scientists of our time."[2]

As a refugee from Europe, Weisskopf's big fear was that the Nazis might triumph in World War II by being the first to launch an atom bomb attack. After all, nuclear fission had been discovered in 1938, a year before Hitler invaded Poland. And Weisskopf's friend, the brilliant Werner Heisenberg, headed the German researchers. If anyone could design such a bomb, thought Weisskopf, it was Heisenberg. To avert this danger he suggested kidnapping him, and even offered to help. Some say an attempt was eventually made to kidnap or kill Heisenberg. But Weisskopf played no role in that if, indeed, there was an attempt. Instead, he worked on the American atomic bomb as a group leader under Hans Bethe, in what Weisskopf saw as a desperate race against the Nazis.

Impressed by Weisskopf's almost uncanny ability to predict the outcome of experiments, colleagues working on the bomb posted this sign on his office: "Los Alamos Oracle." Those less impressed dubbed it "the cave of hot air." Using intuition, which he defines as "imaginative guesswork," he set to work at freeing "for the first time the immense cosmic forces hidden within the atomic nucleus."[3]

Temperatures at the center of the atomic explosion were expected to reach ten times the heat at the sun's core, and the pressures were to be thousands of times greater than any previously produced in a laboratory.

As Weisskopf recalled: "While we were making the two-hour journey by jeep to the test site, Enrico Fermi said it would be a miracle if the atmosphere were ignited. But then he made a jocular remark—in bad taste to my mind—that the chance of a miracle was ten percent." Weisskopf and others correctly took this as a bad joke. "But one of our colleagues took it seriously and, anticipating the ultimate holocaust, had a nervous breakdown."

Although planned for midnight, the test was postponed due to bad weather until five the next morning. Because of that time change, the countdown broadcast over the intercom shared the wavelength with a local radio broadcast of a Tchaikovsky waltz. It was an eerie confluence of the jaunty waltz with the anticipated explosion of the deadliest weapon ever created. Today, Weisskopf never hears that waltz without conjuring up the anxieties of Trinity.

To experience the blast's full effect from ten miles away, he, Fermi, and a few others defied orders to lie flat and face away from the bomb, which exploded with almost three times the force he had expected.

Earlier, when Weisskopf calculated dangerous radiation from the bomb would be harmless at five kilometers from the explosion, colleagues jokingly suggested placing him there in an iron cage. Fortunately he didn't put his theory to the test: it would have killed him.

Thirty-six hours later he drove in a jeep with Fermi and Bethe to measure radiation remaining near the center of the explosion. To

Weisskopf's horror his measuring instrument read almost 200 roentgens—a probably fatal dose. He kept the ominous news to himself, simply warning them that they'd hit a "hot spot" and to run for the jeep. Back at headquarters he quickly found a technician to check the instrument. It was declared in good order. Desperately, Weisskopf reexamined the instrument and found the indicator had been put on backwards. Instead of 200, the correct reading was just over zero. They would live.

I asked Weisskopf: Were you initially afraid the first atomic bomb might set the atmosphere of the world ablaze?

Weisskopf: Very much so. As far as I remember, I personally led a group to do calculations. Fermi helped us. This was taken very seriously. It took at least a year at Los Alamos, and we convinced ourselves that it would not set the world's atmosphere on fire.

Hans Bethe told me your account of calculating if the whole atmosphere might catch fire when the first atom bomb was tested was "a nonsensical idea. And I worked that out. Victor Weisskopf was not much concerned with it."

Weisskopf: Either he or I must be wrong. My memory is not as good as Bethe's, so maybe my account is incorrect. I was also the theorist for the test and I calculated everything that was important for the test. I was about 10 miles, 16 kilometers, from the explosion. And suddenly when it exploded I felt strong heat on my face, which I should have expected. I mean, heat also goes with the velocity of light and should have reached me at the moment I saw the bang. But at that moment, suddenly, my heart sank—for a second—afraid that the atmosphere was burning. But the heat was to be expected.

E equals mc^2 is supposed to have heralded the atomic bomb. But that formula doesn't say anything about *how* to change mass into energy, does it?

Weisskopf: Of course. It's ridiculous to say Einstein made the atomic bomb. What you said is the exact answer to it.

Victor Weisskopf grew up in Vienna, Austria, where he was born on September 19, 1908, the son of cultured and assimilated Jews. He graduated from the University of Vienna, and teamed up with Eugene Wigner in Gottingen, Germany to investigate light emitted by atomic electrons. In the 1930s he worked with two other great scientists, searching for the elusive secrets of the atom. He worked first with Niels Bohr in Copenhagen, then as Wolfgang Pauli's assistant at Einstein's alma mater, Zurich's Federal Institute of Technology.

In 1936 Weisskopf traveled in the Soviet Union "to make my own judgment about the character of the regime and was discouraged by the atmosphere of fear and terror." Two years later with a European war imminent and Nazi savagery against Jews officially sanctioned, Weisskopf arrived in the United States as a refugee. On trips to visit physicist friends from Europe, now also refugees—Bethe at Cornell, Wigner at Princeton, Teller at George Washington University—Weisskopf also called on Robert Oppenheimer at his New Mexico ranch. When they went horseback riding together, he told Oppenheimer of persecutions under Stalin and the terrible conditions in the Soviet Union. And, he believes, these conversations helped to disillusion Oppenheimer in Communism.

After the war Weisskopf was a founder of *The Bulletin for Atomic Scientists*, which advocated nuclear disarmament. In 1961 he became director-general of CERN in Geneva, Switzerland, supervising physicists from fourteen nations seeking to study the fundamental structure of matter. Five years later he returned to America as chairman of Massachusetts Institute of Technology's (MIT) physics department.

A music lover, he celebrated his retirement at MIT by playing the piano in a Beethoven trio and conducting an orchestra in a Bach Brandenburg concerto. Weisskopf was happily married to Ellen Margrete Tvede, a Dane, for 55 years. She died in 1989. Their son, Thomas Emil, is a professor of economics at Michigan University, Ann Arbor. Their daughter, Karen Louise Worth, is involved in science teaching in elementary schools throughout the nation—a project of the National Academy of Sciences.

Today, Weisskopf is active as an emeritus professor of physics at MIT. Mozart and quantum mechanics are still his sustaining passions.

In his autobiography, he gives readers vivid glimpses of his encounters during a life in science with "the best brains, the best people in my profession."

Oppenheimer: "An unusually inspiring leader . . . His ability to be ready with the answer before one had finished formulating the question helped him to be aware of everything interesting that happened on the hill." (The location of Los Alamos)

Fermi: "As the boom (of the test explosion) approached, Fermi stood there, erect and calm, slowly tearing little pieces of paper and letting them fall to the ground. He held a measuring stick in his hands. The displacement of air from the shock waves caused the slips of paper still falling to move horizontally by almost a foot. He measured the distance, looked into his notebook, and said, 'Twenty thousand tons'—an example of Fermi's direct, simple approach to problems of physics. (similar to Feynman's) A more sophisticated instrument could not have measured the efficiency of the explosion more accurately."

Feynman: "Richard Feynman became one of the foremost theoretical physicists of our age. He also added to our group because of his charm, playfulness, and personal warmth. Extremely witty, he was master of the practical joke. For instance, he could program our mechanical calculating machines to produce a rhythmic clatter that mimicked popular songs of the day. To our children, who adored him, he was the funniest man in the world and their favorite adult playmate. Feynman's subsequent contribution to science was enormous. Modern field theory would be unthinkable without his decisive work. He conceived seminal ideas in almost all fields of physics. The only theorist I think of who was comparable to Feynman was Lev Landau."

Landau: "When I visited Lev Landau's bathroom, I found a volume of Stalin's autobiography where the toilet paper should have been. This was typical of Landau, who was famous for taking political risks. Although Nikita Khrushchev had just made his first public statement about Stalin's misdeeds, using Stalin's book for toilet paper would probably not have been considered approved behavior. Unfortunately, an automobile accident in 1962 ended Landau's intellectual life. His brain was damaged in such a way that he could no longer read a book or a paper, much less formulate ideas. He remained in a hospital until his death in 1968. I visited him there and it was a harrowing experience. His short-term memory was destroyed, but he remembered everything about the past . . . Every three minutes or so he forgot that I had been with him for some time and said, 'Oh, how nice of you to visit me here in the hospital.' "

Sakharov: "After I had visited with Sakharov and his wife, Elena Bonner, for two hours, Andre brought me down to the street. He embraced me before I entered the car. As we started off, the driver, obviously a KGB man, said, 'You should know that Sakharov is a very bad man.' I replied, 'How can you say that? He is a dear friend of mine and a very great scientist.' He admonished me, 'You should use your influence to make him stay with science and not dabble into politics where he does immense harm.' I expressed my complete disagreement and my conviction that his political ideas were of great significance and should be followed. I don't remember everything I said, but I am sure it was duly reported word for word and can be found in the KGB files on the academy . . . Sakharov was vindicated at the end of his life. The changes in the Soviet Union under Gorbachev go back to his ideas. I doubt they would have occurred without Sakharov's persistence and courage."[4]

WERNER HEISENBERG

One remaining mystery of World War II is whether or not Werner Heisenberg tried to help or to sabotage Nazi efforts to produce an atom bomb, and if his wartime visit to Bohr, his friend and mentor, in Denmark—then occupied by the Nazis—was with good or evil intent. Few believe Heisenberg to have been pro-Nazi. Some think he was a Nazi pawn. Others see him as a courageous liar who tricked Hitler into thinking the atomic bomb was an impractical dream. Physicist Samuel Goudsmit, an expert on the subject, was convinced that had Heisenberg been able to build the bomb he would have done so.

Weisskopf had known Heisenberg well and believes he "did not comprehend the deep hatred and terrible desperation that permeated Denmark, a nation victimized by the Nazis. (Visiting Bohr in 1942) Heisenberg expressed himself vaguely, fearing that any direct statement about Germany's nuclear effort, or any doubt of German victory, would put him and his family in mortal danger. Under these conditions, it was difficult for Bohr to see Heisenberg as a disciple and friend and not as a representative of the oppressor. For his part,

Heisenberg should have been expected to trust Bohr. He should have assumed that his old friend would have taken every precaution to prevent their being overheard and would never have given him away. An open exchange could thus have been useful, but it did not take place. Consequently, a great friendship was shattered, and a creative human bond was severed ... We know that Heisenberg did what he could to protect the Bohr Institute during the German occupation, and he saved many people's lives ... Before the war he always struck me as an innocent Boy Scout type, free of worries, youthful and enthusiastic. But when I saw him again, (after the war) even his complexion had changed, and this was not due to old age. He visibly carried a load. I could not help thinking of Oscar Wilde's *The Picture of Dorian Gray* when I saw the imprint of those tragic years on his face."[5]

When I first spoke with Victor Weisskopf, Heisenberg was very much on our minds. Two books had recently appeared about Heisenberg, one by Thomas Powers, the other by David Cassidy.[6] I had discussed his book with Powers. And Weisskopf had just arrived at his home in Newton, Massachusetts, from a talk he had given, and where he had encountered the author of the other Heisenberg book, David Cassidy. Cassidy had shown Weisskopf his negative review of Powers' book that was forthcoming in the British journal *Nature*.

So, I said to Weisskopf: I was fascinated by Tom Powers book *Heisenberg's War*, in which he mentions your plan to kidnap Heisenberg. William Stephenson, head of the British Secret Service in the U.S. during World War II, told me of plans to kill Hitler early in the war.

Weisskopf: That would have been a better idea.

I suppose the truth about Heisenberg's role in the war would have been hard to get.

Weisskopf: I don't believe this. When I was director at CERN in the early '60s I had a lot of contact with Heisenberg, though at the time it was sort of taboo to talk about these things. But I got a little impression from him and I think I knew him pretty well.

Did he fake his math or do anything else to go slow on a potential German atom bomb?

Weisskopf: I don't think Heisenberg did that. But he probably would have done it, I think, if he were forced to work on it. Fortunately for him and the world, he was able to convince the chief of staff of the Nazi army that it was impossible to make the bomb within a year or so. (Hitler's deadline)

Do you think he genuinely believed that?

Weisskopf: He genuinely believed it, and rightly so. We took over three years with our tremendous industrial base, and the Germans were really not able to do it.

What do you think Heisenberg was really trying to say during his wartime visit to Bohr?

Weisskopf: That's difficult. Heisenberg was not a good politician or psychologist, and he misjudged Niels Bohr's attitude in an occupied country. This I can say with a little more conviction, because I've talked about this with Heisenberg. He wanted something that was impossible at that time, to have the international society of scientists refuse to work on the atomic bomb.

In other words his visit had a good purpose—to discourage the creation of an atomic bomb.

Weisskopf: And an impossible one.

Exactly. And Bohr misunderstood his motive.

Weisskopf: Bohr did not quite understand because Heisenberg was not very courageous. Having his wife and five children to think of, he didn't want to risk anything. And he didn't express himself very clearly which, I think, was rather silly because he could have relied upon Bohr's not giving him away.

So you continue to admire Heisenberg?

Weisskopf: Yes, except that in this situation he was not courageous enough. There were Germans, for example Helmut von Moltke, who were much more courageous. Moltke did a number of things against the regime, for which he was executed. And there was the famous case not enough people know about, of George Dukwitz in Denmark. He was an employee of the Nazi Embassy, and he was told secretly in the embassy that all the Jews in Denmark were to be transported to a concentration camp. He gave the news to the Danish Underground who were able to save practically all the Jews.

I can understand Heisenberg's patriotism and wanting to stay in Germany with his family to create "islands of decency," or some such phrase he is supposed to have used in discussing the situation with Planck. But if he read *Mein Kampf*, which as an intelligent man I presume he did, I don't see how he could have stayed knowing what Hitler's plans were.

Weisskopf: He despised Hitler. He belonged to that Wednesday Society of people who wanted to assassinate Hitler very much. So he certainly despised Hitler, but he did not despise Germany. And he said, "This will be over some day and I would like to save the good things."

The Dutch physicist, Henrik Casimir, recorded Heisenberg's words when he visited Nazi-occupied Holland on October 18, 1943: "History legitimizes Germany to rule Europe and later the world. Only a nation that rules ruthlessly can maintain itself. Democracy cannot develop sufficient energy to rule Europe. There are, therefore, only two possibilities: Germany and Russia, and perhaps a Europe under German leadership is the lesser evil."

Weisskopf: First of all, Heisenberg was very much afraid of the Communists. He was afraid the Communists would take over the world. So he said it's a smaller evil if Germany would win the war instead of the Communists. A point of view I do not share but you certainly can discuss it. But to say such a thing in a Nazi-occupied country shows how little Heisenberg understood—the same with his visit to Bohr—the psychology of people occupied by the Nazis.

Would you say then that Heisenberg's fear of the Russians made him more tolerant of Nazi behavior?

Weisskopf: No, no, I wouldn't say it. He was certainly not tolerant of how the Nazis treated the Jews.

Einstein, on the other hand, had a greater fear of the Nazis than of Communists, so was a little more understanding of Communists, though not approving of them, than he was of the Nazis.

Weisskopf: So was I.

And Heisenberg was the other way around.

Weisskopf: Correct.

Hans Jensen, also a German physicist, followed Heisenberg in trying to talk with Bohr about work on the atom bomb in Germany. What was actually said is still a matter of conjecture, but Bohr told you that British Intelligence officers convinced him that Jensen was an agent provocateur sent to pull the wool over Bohr's eyes about the Nazi bomb program. Was Bohr also convinced Heisenberg was an agent provocateur?

Weisskopf: Jensen was not well known to Bohr. He didn't belong to the Bohr circle of physicists. So obviously anyone in an occupied country is terribly suspicious of any German who comes. Jensen was absolutely anti-Nazi and very courageous. Jensen told an Underground guy in Norway in contact with the British Secret Service that Germany was not making the bomb. The British Secret Service was convinced of this in 1942. But General Groves (in charge of the Manhattan Project) did not believe the British Secret Service, because it refused to tell the sources, not wanting to endanger them. Groves, whom I didn't like especially, was a very suspicious guy. He thought, only if the British Secret Service gives me the sources then I can believe them.

Sounds like Hoover of the FBI.

Weisskopf: Not quite as bad as Hoover. Because the greatest piece of Groves' work was hiring Oppenheimer [to be scientific director of the Manhattan Project, despite his previous friendships with several Communists].

Thomas Powers writes in *The New York Review of Books*, "What clearly distinguishes Weisskopf's account of Heisenberg's role is generosity. He is willing to make allowances for Heisenberg's refusal to leave Germany in the 1930s." And Jeremy Bernstein in the same magazine is annoyed because a careless reader could think you are applauding German scientists for their moral stand in not enthusiastically supporting atomic bomb research. And, judging by the transcripts, [a result of the British Secret Service bugging Farm Hall, the mansion where unsuspecting German scientists were prisoners] Bernstein believes Heisenberg failed to get the correct dimensions for critical mass because of incompetence, not lack of enthusiasm.

Weisskopf: That's not true. Actually he got it within two weeks when he really tackled it at Farm Hall.

Bernstein also calls Heisenberg, "naive, arrogant, insensitive and egocentric."

Weisskopf: I disagree. Naive, perhaps.

Did Heisenberg work on *anything* of military importance to the Nazis?

Weisskopf: Nothing.

Powers quotes you as saying of Heisenberg, "We never talked [about his wartime work on the atom bomb]. I blame myself for this. I never said to him, 'Tell me what happened.' "

Weisskopf: That's a very interesting thing, which I don't understand. I could kick myself. In the '60s when I had very strong connections with Heisenberg at CERN, I don't understand why I didn't speak with him about these things or with (Fritz) Houtermans, whom I also knew very well and saw a lot of at CERN. I always believed the best thing to keep Germany from again falling into that kind of disease is to bring it into a European framework, therefore I considered CERN very important. Heisenberg was there as a representative of the German government and also as a physicist interested in the subject.

I understand he felt isolated there, and at meals sat alone at a cafeteria table.

Weisskopf: I'm very doubtful that's true. You see at CERN, it was in the '60s, it was long after the war. Sure, there were a lot of people who might fight over the Heisenberg issue. Many thought he was a Nazi and wanted the atom bomb for Hitler. Sure, there were some people who would have shunned Heisenberg. But if he sat alone in the cafeteria it can only have been one reason: that people in Europe are very personality conscious and a young fellow would think it's assuming if you sit with Heisenberg. But I never saw Heisenberg alone. The idea is a little ridiculous.

Heisenberg reminded you of Dorian Gray. Did others remark how he'd aged very fast?

Weisskopf: Yes.

And you think that was because of the great conflict he had with regard to the atom bomb?

Weisskopf: Certainly.

After World War II did you ever mention to Heisenberg your plan to kidnap him?

Weisskopf: No. And I think this is a tremendously overblown story. Indeed I was very mad that someone published it, because it was just a cocktail party idea and it was refused by the U.S. Secret Service.

Elsewhere you called it "harebrained" and and "ill-conceived." But at that time you really feared Germany would get and use the bomb before we did. So your suggestion was to kidnap one man to win the war and save possibly millions of lives. It seems a great idea to me. Maybe Heisenberg would even have welcomed it if he was so anti-Nazi. Wouldn't his kidnapping have been a small price to pay for saving democracy and so many lives?

Weisskopf: Yes, but it would have been dangerous. The people who arrested him would have been shot. Probably Heisenberg would have been surrounded by Nazi guards if he really were the leader of some important war project, which he wasn't.

The most you did, if I've got it right, was to suggest you would point out Heisenberg to American agents when he was lecturing in Switzerland. Then the agents would try to talk him into either not returning to Germany or into coming to America. And to kidnap him if talk didn't work.

Weisskopf: Something like that. I wasn't even sure I'd go over with them. Then in 1944 when they sent Moe Berg [a professional baseball player recruited by the OSS], I had no idea of it and it had nothing to do with me.

But your suggestion may have caused Moe Berg to go.

Weisskopf: Maybe.

Apparently Moe Berg was prepared to kill Heisenberg.

Weisskopf: I'm not sure I believe this 100 percent. Nor would I attack Tom Powers for writing this [in *Heisenberg's War*]. He's been attacked by so many other people—David Cassidy, for example, who wrote *Uncertainty*, [a biography of Heisenberg], which I don't find so good. Yesterday Cassidy was at a talk I gave and showed me his review of Powers' book. I find it very bad taste to completely criticize one's competitor.

I discussed Powers' book with him recently and he said, "In my opinion, Heisenberg did what he could to ensure that there would be no German bomb. There was no throwing of spanners into the works so far as I know, but when you examine what Heisenberg did and said to different people over the course of the war, it becomes increasingly difficult to interpret his advice to German authorities as objective and disinterested. It becomes increasingly clear that he was deliberately telling them things that would discourage them from going forward. Weisskopf has probably the friendliest attitude toward Heisenberg on this question among any of the principle scientists who were involved. Cassidy has a deep-seated and visceral distrust of Heisenberg on this question for reasons I can't understand. And, in my opinion, it says more about Cassidy than about Heisenberg. I don't know exactly what it says about Cassidy. He refuses to recognize certain facts or to incorporate them into his thinking. He is just not prepared to accept the possibility that Heisenberg deliberately refrained from building the bomb. The question to me is an important one and an enduring one." I also read David Cassidy's critique of Power's book, *Heisenberg's War: The Secret History of the German Bomb*, which he called superficial, prejudiced, and quite unconvincing. He also complained that Powers largely ignored the vast paper record, to provide, quote, "a shadow history of what German scientists thought, felt, and said to each other in the small hours of the night about the work they had been given to do."[7] And that Powers was too partial in accepting Heisenberg's and Weizsacker's words, while discounting the views of Bohr and Goudsmit. Cassidy wrote: "Heisenberg's main concern for years had been the preservation of German science, not scruples, and wartime nuclear research provided a new opportunity to pursue the goal . . . There is no indication that Heisenberg delayed nuclear research during 1941 and 1942, although he did express relief in the end that it did not lead to an atomic bomb. Rather than dragging his feet, he worked intensively on two nuclear projects, in Leipzig and Berlin, until June 1942, even neglecting his family and his personal research to do so."[8] End quote. It comes down to this: Cassidy believes Heisenberg didn't know how to make the bomb and was reluctant to try. Powers disagrees.

Weisskopf: I also disagree with Cassidy. There's one thing about which historians of science make mistakes. They accept only what is written, and not the spoken word: it's a kind of pseudoscientific attitude. For example, they even criticize Pais' book on Einstein, ("*Subtle is the Lord*"), which I consider one of the best books ever written. They don't accept these personal impressions.

How impressed were you with James Franck's reply to Heisenberg's wife in the 1950s when she complained that people treated her and Heisenberg coldly and they felt isolated. And Franck said, "This is the way we Jews were always treated—now the Germans must live with it."* A pretty fair comment?

Weisskopf: Very fair.

Isn't the comment of Bohr's widow convincing evidence that Heisenberg's visit to Bohr during the war wasn't humanitarian? According to Powers' book, she said to Goudsmit while pointing at Heisenberg and Weizsacker, "Goudsmit, that wartime visit of those two was a hostile visit, no matter what people say or write about it."

Weisskopf: I disagree very much with the Bohr family. You know my veneration for Niels and also for Margrete (his wife). We were great friends—my wife, who was Danish, especially. First of all Bohr was not a good listener, and being in an occupied country (he would be suspicious of Heisenberg's motives). You know, of course, about the rescue of the Danish Jews. I feel deeply honored because I've been asked by the Danes to come (October 1993) to celebrate the rescue fifty years ago. I'm to give a talk in Copenhagen. So I'm an honorary Dane.

---

* James Franck shared a 1925 Nobel prize with Gustav Hertz for showing that atoms in collision gain or lose energy in quantum steps. In 1935 he left Europe for the U.S., working first at Johns Hopkins and later the University of Chicago.

[After this conversation with Weisskopf, I spoke with historian Mark Walker, having read the manuscript of his book *Nazi Science: Myth, Truth, and the German Atomic Bomb*. Walker had gathered material from German and American archives and interviewed two dozen German scientists who worked with and around Heisenberg. He said: "I looked at Heisenberg's two reports in 1939 and 1940 on the theoretical underpinnings of nuclear physics, which Heisenberg signed and dated, and they include such things as pointing out that pure uranium 235 is a nuclear explosive. And these reports were sent to the army. In fact, the army stamped them the day they got to the office." In his book Walker concluded that in a 1970 private letter Heisenberg made "false" and "tragically absurd" claims that "together with Hahn and von Laue, [he] had supposedly falsified the mathematical calculations in order to deny nuclear weapons to Hitler . . . Heisenberg may have resisted Hitler, in his own mind. [But] Heisenberg's behavior was not so different from most of his colleagues in Germany, the United States, or the Soviet Union who worked on nuclear fission. Almost all of them cooperated with their governments under very different conditions, either out of conviction, ambition, or fear. There was an important difference, but that lay with the political, ideological, and moral nature of the regime, not the scientist."[9] After speaking with Walker I got back to Weisskopf.]

Here's what historian Mark Walker says, "It's not that Heisenberg was so bad. But he was no angel."

Weisskopf: I never said he was.

"It's not that he willingly worked for Hitler. It's just that he was not a resistance fighter. He was in the middle and he was troubled about what he was doing. But he did it. As for that circle he belonged to, some of whose members attempted to assassinate Hitler, Walker said, "It always has a few scientists and cultural leaders, but mainly the circle was connected with the Prussian officer class, who also had an interest in cultural matters. By all accounts of those in the circle who survived, it was anti-Hitler but pro-German: they very much wanted Germany to win the war. And all the people who survived made it quite clear they were shocked and surprised when they heard of the attempt on Hitler's life. It was not an assassination club where everyone knows that so-and-so is planning the bomb attack. Heisenberg and others were utterly stunned." One other thing Walker said:

"It is clear that Heisenberg expected Germany to win the war when he visited Bohr in Denmark and advised him to cooperate with the German authorities." Finally, Walker said that even the much admired von Laue had to make small compromises with the Nazi regime to keep his position as a professor. Do you agree?

Weisskopf: Yes.

I also learned from Walker that there are letters in the National Archives from you to Goudsmit shortly after the war in which you were much less sympathetic to Heisenberg than you are today.

Weisskopf: That's right.

Why?

Weisskopf: Because I found out more about him and his attitude and so I found he wasn't as bad as his critics said.

I noticed in the Farm Hall transcripts none of the German scientists said something like, "We couldn't make the bomb because we got rid of some of our best brains—the Jews."

Weisskopf: But I don't think that's true. [That getting rid of Jews was the reason they couldn't make the atomic bomb.] They could have said it and I would be glad if they did.

You don't think they might have developed the bomb if, for example, you, Szilard, and Wigner had stayed?

Weisskopf: Szilard is the only one. Let me say it this way. To make a bomb is not such a difficult thing scientifically. It is really a technical, not a scientific, problem. And Jewish refugees were not so much the technicians and engineers. Although I must say it's interesting how we physicists, not only refugees, think Germany had a number of people who could do it, and the knowledge was essentially there. But to say Germans didn't get the bomb because they expelled Jews is vastly exaggerated.

The only German scientist Einstein respected after the war was Max von Laue. Would you add any more?

Weisskopf: Among the scientists I'd certainly add Jensen and, of course, Wolfgang Gentner. He was a hero. An anti-Nazi. Einstein might not have known Gentner. I personally would have added Heisenberg.

TELLER AND OPPENHEIMER

Do you despise Teller for his testimony against Oppenheimer?

Weisskopf: Yes.

What do you think his motive was?

Weisskopf: I'm not a psychologist. I saw a lot of Teller in Los Alamos, and the personal relations between Teller and Oppenheimer were terrible. Teller always had great ideas he wanted to apply right away toward the hydrogen bomb and not waste time with the atomic bomb, which was certainly wrong. And finally he got a group to work with the hydrogen bomb. He was dissatisfied with many of Oppie's decisions. For example, when Oppie made me deputy chairman of the theoretical division, of which Bethe was chairman, Teller was very disagreeable. [Teller believed he was more qualified for the job; Weisskopf believed he could get along with people better.]

George Wald's impression of Oppenheimer was that if someone presented a point of view with which Oppenheimer agreed, he'd immediately change his view and oppose them. And he was amazed to see Oppenheimer, whom he considered vain and arrogant, in Bohr's company acting as if Bohr was his god "almost trying to work his way into Bohr's skin."

Weisskopf: The first remark when someone presented Oppenheimer's view he changed his policies is completely wrong. You know, I never take George Wald 100 percent seriously. Oppenheimer's admiration for Bohr is true and I share it. The way Wald described it seems a bit exaggerated, but it's certainly true.

Did you know Oppenheimer's wife, Kitty, was a niece of the German general, Keitel?

Weisskopf: Yes.

Didn't that worry the security people at Los Alamos?

Weisskopf: The point was that she was very leftist and had much stronger connections to the Communists than Oppie had. This probably convinced the Secret Service that her connection with the Germans was not going to influence him.

HYDROGEN BOMB

Weisskopf was one of 12 scientists who stated: "We believe that no nation has the right to use such a bomb no matter how righteous the cause. This bomb is no longer a weapon of war, but a means of extermination of whole populations. Its use would be a betrayal of all standards of morality and of Christian civilization itself." Weisskopf thought he had persuaded Hans Bethe not to join Teller in building the hydrogen bomb, but in 1951 Teller and Ulam found a comparatively simple way to build it, making virtually certain the Soviet Union would also be able to make one. That persuaded Bethe to help Teller. That same year the Korean War erupted. Both superpowers succeeded in building the bomb.

I said to Weisskopf: After World War II you were very outspoken about arms control and against building the hydrogen bomb. But didn't the existence of the atom bomb prevent war between the West and the Soviet Union? To that extent don't you think atom bombs have value in preventing world war?

Weisskopf: Yes, but not in these numbers. A few hundreds are enough. To build 50,000 is a mental disease.

What happened to the hydrogen bomb? Is it stockpiled?

Weisskopf: Sure. The very interesting thing is that Mr. Fuchs told the Russians the secrets, but he told them Teller's idea to build the hydrogen bomb—and it was the wrong one. [Teller, helped by Stanislaw Ulam, eventually found a relatively simple way to make the weapon.] And the Russians were very amused because Sakharov knew right away and built the right one. They built it three years after they built the A-bomb.

How much more powerful is it than the Hiroshima bomb?

Weisskopf: That depends on the size. The Russians made it tremendously powerful.

When you returned to Germanv after the war to see your former teacher, Professor Steppan, you told him you were grateful he had never allowed his anti-Semitism to come between you when he was teaching. And he wept. Did you suspect he was still an anti-Semite despite his tears?

Weisskopf: No. He was one of the few Nazis I knew—I didn't know many—who regretted having been a Nazi.

Among your colleagues, was Lev Landau with his frequent cries of, "Say something; I want to contest it," the most challenging? Weren't several others equally provocative?

Weisskopf: Szilard, perhaps. Pauli wasn't aware of his aggressiveness. His was a kind of deep honesty. He really thought that way and it was a wonderful thing to work with him; a man who expresses exactly what he thinks.

Wasn't Einstein the same?

Weisskopf: But in a more general way, not immediate.

Did you ever hear Einstein play the violin?

Weisskopf: I played with him.

Was he any good?

Weisskopf: Not as good as his physics. He was a real amateur. My piano playing was better than his fiddle.

When you heard Einstein and Bohr discuss quantum theory, who came out the winner?

Weisskopf: I know quantum theory comes out the winner.

But Bohr was very hard to understand because of the way he spoke and Einstein was very clear. Did Einstein win by being more articulate, though he might not have won factually?

Weisskopf: What do you understand by win?

For example, Abraham Pais told me that when a Bohr disciple came to Princeton and argued with Einstein about quantum theory, Einstein definitely won the argument.

Weisskopf: Bohr had trouble in making himself clear and in following the arguments of somebody else. But my admiration for Bohr is known to you. Winning is a funny expression. I can imagine a disciple of Bohr who is not extremely good in understanding the basics of quantum mechanics, which are difficult, would lose against the great spirit of Einstein. But Bohr did not lose. Bohr always had counterarguments. Einstein came with seemingly clear and binding arguments against quantum mechanics, and Bohr could always show Einstein's arguments were wrong.

Did Einstein ever have the better of Bohr in debate?

Weisskopf: Sometimes, perhaps. I find college debates on scientific matters absolutely ridiculous. Debates about political matters I am very interested in.

Pais thought Bohr was a finer human being than Einstein.

Weisskopf: He was, in a way, but it's a one-dimensional view. To me and Abraham Pais, a very close friend of mine, Bohr was a tremendous personality with much deeper philosophical interests and insight than Einstein. But Einstein, of course, was the greatest physicist that ever lived. I mean (he chuckled) Einstein was such a wonderful person. So was Bohr and we need both types.

Are there any living scientists the caliber of Einstein, Bohr, or Heisenberg?

Weisskopf: No, and that worries me: the lack of personalities in our culture.

In what order would you rank the three of them?

Weisskopf: It's very hard to say because again, one should never think one-dimensionally. The fact that Einstein proposed his two theories almost single-handedly, almost without help—Hermann Weyl helped him a little—is unique. I think it's hardly ever happened, except perhaps with Newton. In this respect Einstein is certainly great. Niels Bohr never worked alone, so that's different. Heisenberg was *fantastic*, absolutely incredible, in some ways more than Einstein. Einstein created something new, but so did Heisenberg. In almost every field of quantum mechanics Heisenberg had an essential seminal contribution.

You say Einstein worked alone. How about his close friends like Besso?

Weisskopf: They didn't help in the technicalities of his research.

I thought he wasn't strong in math and they helped him.

Weisskopf: That's not true.

Who at his height would you employ for a difficult scientific problem, Feynman or Heisenberg?

Weisskopf: Now look, there's a tremendous age difference.

But you knew them both in their prime.

Weisskopf: It depends what they have to solve. Heisenberg is probably the greater physicist. He was unique.

Would you agree that Gell-Mann and Feynman were the two leading minds of their generation?

Weisskopf: Yes, but Feynman was greater than Gell-Mann.

CREATION OF THE UNIVERSE

What do you think of the ideas expressed by some science writers that a revolution is taking place in the study of the cosmos? They suggest that the universe evolved from virtually nothing and that the entire universe may be embedded in another, far larger but unobservable universe. And they replace the "Big Bang" theory with an "inflationary universe" model.

Weisskopf: First of all, the inflationary universe is not in contradiction to the Big Bang. Journalists give the impression of always wanting excitement and revolution. It's usually not true. Rather it is evolutionary development of ideas, with very few exceptions like Einstein's and quantum mechanics.

Yet Einstein claimed his views were not revolutionary.

Weisskopf: In a way, he's right. Einstein is actually the crowning of classical physics, whereas the real revolutionary theory is quantum mechanics. I'm not sure the inflationary universe theory is true but it's the idea of Alan Guth (among others) who's here at MIT. This is it: Space is never quiet. That is sort of quantum mechanics: that nothing is completely at rest. There are always fluctuations, field fluctuations, energy fluctuations and so on. And supposedly, very rarely, there's one fluctuation that is rather big. And according to the math of relativity, if you have such a big fluctuation—big doesn't mean big in space but big in intensity, in energy, in density—it means high energy in a small space that has a tendency to expand very rapidly. So there is an enormous inflation very fast. This inflation stops, however, at a certain size. And then a field of energy forms into particles, let's say quarks, guons, and protons, and God knows what. And then I would say the ordinary story of the universe begins. Then it's already pretty big, not as big as now.* Then the ordinary expansion begins, which is much slower than this first explosion. Then it goes on, quarks, and antiquarks, electrons, antielectrons, and so on. And the quarks assemble to protons and neutrons, and the latter to nuclei. The story of the universe. The ordinary story of the universe. I call it ordinary because this rapid inflation which I described to you I call extraordinary. That's just my word for it.

Alan Guth may not agree with all the details I've told you, but it's his idea about inflation and so on. It shows that the creation of the universe may not be a unique event. Before or after that, even other universes may have been created. This has something to do with the statement by Johnny Wheeler. We may not be the only universe. An interaction between different universes is not impossible, but it's extremely improbable.

Is the inflationary theory saying that the Big Bang occurred in two distinct stages?

Weisskopf: I think the inflationary theory IS the Big Bang.

---

* Alan Guth believes our universe inflated from a size smaller than the elementary particles in atoms to many orders of magnitude greater, instantly transforming it from the ultrasmall, following the rules of quantum mechanics, to the more predictable and visible universe we know that's ruled by Newtonian physics.

FASTER THAN LIGHT?

John Bell spoke of two particles separating and seemingly able to communicate instantly over vast distances—faster than the speed of light. And when one particle turned, the other immediately turned as if it was somehow in instant touch, almost "aware" of what the other particle was doing so far away. And Bell concluded: "Separate aspects of the universe may be connected at a fundamental level and that once connected remain attached over distances by some unknown force moving at more than the speed of light." According to John Wheeler, Bell's theory "deals with worlds that never were and never can be." Is this then fantasy, or science fiction?

Weisskopf: I personally have a very high respect for John Wheeler. For all that he did in physics I would have given him the Nobel prize, which he doesn't have. But I do not like the way he speaks about quantum mechanics. (Both Weisskopf and Wheeler knew Bohr and Einstein.) It's very misleading, especially for the noninitiated. Bell's is an extreme case well-known in quantum mechanics. Quantum mechanics is not local. For example, the hydrogen atom is spread out, and obviously if you locate the electron at one edge of the atom it has disappeared at the other side. So that is a kind of action at a distance. I wouldn't call it that, but it gives the impression of that kind of thing. The quantum state is spread out, and the famous Einstein–Podolsky state is simply an extreme case of this, invented because of its extremeness. But it turns out that this is actually so. There were experiments made. It means that in quantum mechanics the state is not localized and it can, under certain very difficult conditions, spread out over relatively large distances. It's misleading to say it moves faster than the speed of light.

But where's the action in your description of the hydrogen atom? Bell was speaking of one particle immediately imitating the action of another when they were separated by great distances.

Weisskopf: There's no action there, either, because it's an expanded state which you recognize on one side and conclude from this what's on the other. So they are not imitating each other but acting as a single unit.

You stress the exactness found through quantum theory. Others harp on Heisenberg's uncertainty principle. Is your point that because everything consists of specific quanta only certain sizes and orbits are possible?

Weisskopf: In my view Heisenberg's uncertainty principle should be called "limiting relations," because it says you cannot apply classical concepts to very small objects. There is a limit to here and no further, that you can apply to our well-known classical concepts like position and velocity. Under these limits there are new concepts, namely quantum mechanics.

Is it correct to describe the quantum as minute?

Weisskopf: No. It depends. First of all, "What is the size of the quantum?" doesn't make much sense.

How does it spread over the large distance Bell talked about?

Weisskopf: The point is there is no such thing as *the* quantum. The size, that is energy, is sometimes minute but not the spatial size.

What do you think of superstring theory?

Weisskopf: I'm skeptical. It's appropriate to my previous remarks about publicity and science. You people always want to get the most exciting news, but as science develops one makes a lot of tries that turn out to be wrong and too difficult. And that's the case with superstrings. There are more important things to transmit to the public like the Big Bang or quantum mechanics itself. Strings theory is one of those fancy ideas that a few people support. I think it has nothing to do with reality, but that's my personal opinion and I may be too conservative and old.

Is there any chance of Einstein's unified field theory being achieved in some way?

Weisskopf: It depends what you mean by "in some way." There's a lot of progress, but first of all we don't have a unified field theory. The unified field theory that's attempted at present is somewhat different from Einstein's, because it contains a lot of quantum theory. So it's hard to tell. I cannot answer you with yes or no.

And you don't think superstring theory is the most promising path to unification?*

---

* John Wheeler discusses superstring theory in Chapter Six.

Weisskopf: Not to me it isn't.

Your former student, Murray Gell-Mann, is enthusiastic about it.

Weisskopf: I know.

He thought some of my questions were stupid. You've taken them to be reasonable.

Weisskopf: Now look here, I have an old principle: stupid questions are very often the most interesting ones.

Somebody else said, "There are no stupid questions, only stupid answers."

Weisskopf: That's very good. I will use this quotation.

As relativity and then quantum theory were the two greatest scientific breakthroughs in the 20th century, do you see anything as vital on the horizon?

Weisskopf: I don't.

I was interested to learn that the pope subscribes to evolution theory and discredits creationism. According to Stephen Hawking the pope said in 1981 that scientists shouldn't inquire into the Big Bang as it was the moment of creation and therefore an act of God. How do you like that?

Weisskopf: I don't.

What do you think of Hawking's suggestion that there was no moment of creation?

Weisskopf: To tell you honestly, I don't understand. The way I described it to you with the fluctuating, that would have been the moment, you know.

CONSCIOUSNESS

Do you find consciousness interesting?

Weisskopf: Tremendously.

Can quantum theory explain it?

Weisskopf: Absolutely not. It's extremely interesting that I convinced my grandson, who's an extremely gifted scientist, now 26, to go into neuroscience.

George Wald believes the entire universe is suffused with consciousness.

Weisskopf: I think this is partially without content. We don't know what consciousness is. It has something to do with the process of thinking and associating, and we have practically no idea what is going on there. I'm personally convinced that animals also have consciousness. Anybody with a cat or dog would agree. But to ascribe consciousness to dead matter seems to me a philosophical idea without any real content.

I think he'd say it wasn't dead matter.

Weisskopf: Sure, sure. Some of Pauli's statements are also interpreted—but I'm not sure he said it that way—to say that dead matter also has consciousness. It doesn't make sense.

Here's a naive question: Is everything made from electricity?

Weisskopf: No, although electricity plays a very important part. Take an atom. It consists of a nucleus and of electrons. In an electric current, it's the electrons that are moving. But the nucleus is charged. This, too, has an electric charge. But it is made of protons and neutrons, and the protons and neutrons are supposed to be made of quarks (one of the fundamental, indestructible particles discovered and named by Murray Gell-Mann). Quarks also carry a charge. But the force that keeps the quarks together, keeps the protons and neutrons together, is the nuclear force, which is not an electric force.

Are the six varieties of quarks regarded now as the basic building blocks of nature?

Weisskopf: Yes.

Is it assumed then that we've discovered nature's absolute building blocks?

Weisskopf: No. Because you should not use the word absolute in physics. That is what we now consider, now have reached, so to speak, as the basic blocks. There's no way to know at present whether the quarks themselves do not consist of infraquarks or whatever you call them. Now let's not forget the electron. The electron is not made of quarks and the atomic nucleus is extremely important for the structure of the atom. Without the atomic nucleus there'd be no atoms. And what holds the atomic nucleus together is the nuclear force. There is now a theory called chromodynamics which describes that force—and that's not electric.

Max Planck gave you misinformation about light as seen as a series of particles. Is the electron the only particle that can travel at the speed of light?

Weisskopf: No. An electron cannot travel at the speed of light. Anything that has mass cannot travel at the speed of light. But in science nothing is absolutely permanent. We have no dogma. Although that light velocity is the limit is most, most, most probable.

Is the light quantum (photon) the only particle that has no rest mass?

Weisskopf: No, it's the best known one. Supposedly gluons have no rest mass either. And neutrinos might not have rest mass, therefore they travel with the speed of light.

Here's a simple question that's bothered me for years. The earth spins because there's nothing to stop it, am I right?

Weisskopf: In the case of Venus, (which doesn't spin) probably solar attraction prevents the spin. And I wouldn't say we'd spin forever.

But does the earth keep spinning because there's no friction?

Weisskopf: Let's not forget there's a little friction because space is not completely empty. It's full of gases and things. It's extremely little, but enough to cause a little friction.

Then the earth's spin is slowing?

Weisskopf: It is slowing. Also the tides are slowing it.

What started it spinning?

Weisskopf: There are theories that are not too reliable. But the idea is that gas that surrounded the sun—it's pretty sure because we observe it in other stars that there is a kind of flat ring around a star at the beginning, not unlike the ring around Saturn. And then within this ring of material, units are formed by gravitation, and these are the planets. Since the gas has to run around the star, because if it didn't it would be absorbed into the star, it has an angular momentum and somehow this angular momentum shows up somewhere. And that's the rotation of the planet.

The spinning has nothing to do with the Big Bang?

Weisskopf: Absolutely not.

In 1960 you said you were inclined toward continuous creation [The expansion of the universe is accomplished by continuous creation of matter in space. Whenever the expansion leaves an abnormally large space between galaxies new stars and galaxies are created— out of nothing.] Have you changed your mind in the past thirty odd years?

Weisskopf: In the first edition of my book, *Knowledge and Wonder*, I was rather impressed by continuous creation but then I regretted that, mainly because of the famous three degree radiation—that space all over is full of radiation from the reverberations of the Big Bang— which made me decide otherwise.* I had been impressed by continuous creation because it relieved us from God in some way. A very few people still believe in it.

Galaxies appear to be moving away as if from the Big Bang, is that right?

Weisskopf: Away from each other.

But aren't their movements caused by the Big Bang?

Weisskopf: That we don't know. You used the word "caused." Nobody really knows the cause of the expansion.

Can you say the expansion followed the Big Bang?

Weisskopf: Followed in time is certainly true.

And what part does the so-called Super Attracter play?

Weisskopf: Galaxies tend to be moving toward one region, so one says there must be a Super Attracter.

Do you believe in the likelihood of intelligent life on other planets?

* Discovered by Robert Wilson and Arno Penzias. See Chapter Eight on Penzias.

Weisskopf: Ernst Mayr, a very good friend of mine, said and I very much agree with him, that we have overemphasized intelligence. There's a famous joke: "You look for intelligent life on other planets? Better look for it on our planet." But never mind, that's another story. You see, life on our planet had a certain development. Darwin's hypothesis of evolution describes and explains such development from the most primitive cell to humankind. Somehow it turned out that the ability to think, construct, and build had an extreme value for sustaining life—and therefore developed. Now there's no reason whatsoever that if the whole life evolution is repeated that it again will produce the scientific–technical culture. I always say science is important but it's only part of our reaction to our environment. There are others: morality, religion. And the idea that any evolution must end up with technical prowess seems to me very improbable. I don't doubt that there's life somewhere in the universe. After all, there are billions of planets and possibilities of life like here, but there's no reason it should end up with scientific–technical achievements. Therefore I think it's pretty hopeless to look for another technical humanity.

Harvard physicist Paul Horowitz says: "In the Milky Way, where we live, there are at least 100 billion suns. Within range of current detection instruments are 10 billion other galaxies, many larger than our own. Beyond them may be gigantic numbers of galactic islands— perhaps an infinite number, and corresponding to that, an infinity of suns. Intuition would seem to demand that in such a vast firmament many hearts would beat. I feel nearly certain the galaxy will turn out to be rich with life, of many shapes and sizes, and that any civilization we contact will be far wiser than we. To think we are the best the universe could manage—the mediocrity of it all!"[10]

Weisskopf: I agree. If you read the newspaper every day (He reads *The New York Times* and watches TV's MacNeill-Lehrer.) you see that the race of man is not a very well-designed thing.

Yugoslavia, perhaps.

Weisskopf: For an example. The Nazis are another.

Back to your subject. Are you disappointed that research is no longer done by individuals but in large groups?

Weisskopf: That isn't quite true. These big groups always have one man who's the leading spirit.

In *Genius*, the biography of Richard Feynman, you're quoted as saying: "The theory of the elementary particles has reached an impasse. Everyone has had enough of knocking a sore head against the same old wall." "Is that why you persuaded your grandson to study neuroscience rather than physics?

Weisskopf: No, not at all. On the contrary, that would be a reason to stay in physics. But it's no longer such an important problem for human existence as the brain, which I'd study if I were younger.

So you're not saying physics is unimportant?

Weisskopf: Everything is important, but there are degrees of importance and of personal interest.

But brain research is more challenging?

Weisskopf: Yes, because it's completely unsolved and the problem of the atom and so on are to a great extent solved.

You wanted your son and daughter to carry on your ideals. Weren't you worried they might grow up to scorn your ideals?

Weisskopf: Sure I was worried, but I give credit to my wonderful wife.

And kept your fingers crossed.

Weisskopf: But it turned out extremely positive.

Do you think humans are a vital part of nature?

Weisskopf: What do you mean by vital?

Another way of putting it: Do you think there's a purpose to life, that humans are on earth for some purpose? John Wheeler thinks the purpose is to ask questions. To question—and discover the answers.

Weisskopf: That's a little poetic. Bohr says—it's in German so I'll have to translate—"the meaning of life consists in the fact that it makes no sense to say life has no meaning."

I'm puzzled by Bohr. Once when asked whether something existed he answered, "What do you mean by exists?" As if by questioning your question he had answered it.

Weisskopf: Or he says to a fellow, "Think about it better." I think life has a meaning. Some people don't see any meaning with bad consequences to their psyche. I think the whole development of the universe is a miracle. I don't mean that I believe there's a God who invented it. It's a miracle that we are here. The fact that we can enjoy the spring. This is a miracle and a pleasure. And music! Yesterday I was at a concert of Beethoven sonatas. That such fantastic masterpieces were made by humans makes life worth living.

RELIGION AND MYSTICISM

I'm paraphrasing from your book, *The Joy of Insight*, now. When Wolfgang Pauli was suffering from severe depression, he went to Jung for treatment which led him to become interested in Jewish mysticism and to a friendship with Gersholm Scholem, the world's greatest authority on the cabala. Pauli eventually urged you to visit Scholem when you were in Israel and to go to the tomb at Sfad of Rabi Isaac ben Solomon Luria, an eminent interpreter of the cabala. You went to the cemetery with Israeli mathematician Chaim Pekeris on a Saturday, even though traditionally it's forbidden to visit the cemetery on that day. But to show respect, you and Pekeris wore skullcaps. As you walked to the grave, although it had been a clear day with no wind, a sudden gust of wind blew off your skullcap. You searched for twenty minutes but couldn't find it. Both of you were a little shaken but you persuaded Pekeris to continue to the tomb. On your return you both again searched for the skullcap in vain. In your book you wrote, "It had miraculously vanished in the wind," and you concluded that, "All mystical and psychic experiences should be considered complementary to the scientific approach." Pauli's interest in these different avenues of human experience was in many respects a natural expansion of his involvement in modern physics.[12] Do such subjects interest you?

Weisskopf: I'm interested in mysticism as a human phenomenon. When I lost that skullcap, I told Pauli's friend, Gersholm Scholem, [whom he met the next day], and he said, "Ah, I'm glad that you as a physicist had a brush with the irrational."

Was Pauli the only physicist you knew interested in mysticism and the extrasensory?

Weisskopf: No. Fiertz, for example, who was influenced by Pauli.

Does telepathy interest you?

Weisskopf: Yes, everything interests me. But I think one must be very careful. There's a lot of cheating. I've always taken an interest in things that are not scientific, because it's my fundamental view that science is only one way of dealing with human experience, and if you concentrate on one you are getting into trouble. In the Middle Ages, it was religion.

Any interest in hypnosis?

Weisskopf: It's an interesting phenomenon.

Do you know any great physicists who are deeply religious?

Weisskopf: Yes, Arthur Compton is one.

How about your contemporaries?

Weisskopf: No.

Do you believe Jesus Christ was a historical figure?

Weisskopf: Yes.

I take it you don't anticipate life after death?

Weisskopf: No.

You don't believe in some superior intelligence behind it all?

Weisskopf: That's too specific. I believe there is something that should be in people's minds which is great and awe-inspiring. Call it God or divine. Certainly there's something divine in people who do good.

Chapter Five.   Hans Bethe

Hans Bethe "One of Nazi Germany's greatest gift to the United States."
—*Time Magazine*

On the one hand Hans Bethe is reassuring. The fusion of hydrogen in the sun, he says, will be a source of life-sustaining energy for our earth for billions of years to come. He worked that out. On the other hand, working on almost the same principle as the sun, he knows the hydrogen bomb in the wrong hands could end life on earth.

He was head of the theoretical division at Los Alamos, which produced all the vital ideas on how to make the atomic bomb. But he scoffs at the suggestion that there was any danger of a chain reaction run wild setting the atmosphere of the world alight. Even though one of his men had a nervous breakdown at the prospect. And at the time, Edward Teller's calculations indicated that a fusion explosion might set the nitrogen in the earth's atmosphere afire. Oppenheimer had immediately confided to his colleague, Arthur Compton, "Found something very disturbing, dangerously disturbing . . . No, not to be mentioned over the telephone." Compton was so shocked when Oppenheimer spelled out the danger that they considered abandoning the project. As Compton said, "Better to be a slave under the Nazi heel than to draw down the final curtain on humanity." The chance of ending the world in a holocaust was later recalculated as three in a million, which Teller translated into, "It couldn't happen."[1] And the project continued.

EARLY CAREER

In 1929 when he was 23 Hans Bethe showed signs of promise as a physics instructor. At Stuttgart University he introduced math methods that have been adopted by many researchers in molecular structure. Between 1930 and 1932 he worked with Ernest Rutherford at Cambridge and with Enrico Fermi in Rome, before returning to lecturing at the Universities of Munich and Tubingen. But in 1933 the Nazis took over and, because his mother was Jewish, Hans was kicked out of both jobs. Welcomed in England, he taught at the University of Manchester and was a fellow at the University of Bristol. And in 1935 he accepted an offer from Cornell to be an assistant professor of physics.

In six weeks during 1938 he worked out what made the sun shine, confirming Einstein's $E = mc^2$: the enormous amount of energy produced by the conversion of hydrogen into helium equals the mass of the consumed fuel times the square of the velocity of light.

He became a naturalized U.S. citizen in 1941, and between 1942 and 1946 headed the Theoretical Physics Division of the atomic bomb factory laboratory at Los Alamos. Among those working under his direction were Edward Teller, Leo Szilard, Richard Feynman, Victor Weisskopf, and a scientist who later turned out to be a spy, Klaus Fuchs.

Bethe was awarded the 1967 Nobel prize "for his contributions to the theory of nuclear reactions, especially his discoveries concerning the energy production in stars." He was a professor of physics at Cornell until 1975, and since then has been a professor emeritus there.

I spoke with the 87-year-old Bethe, still active and engaged, on May 12, 1993.

How close could a spaceship get to the sun on its 93-million mile journey before being pulled into its gravity?

Bethe: Long before that certainly humans would be burned up. I've never thought of that question. But I think it would be dangerous to get to the distance of Mercury (the nearest planet to the sun, which on its orbit gets as close as 28 million miles and as far away as 43 million).

Your friend Theodore Taylor planned what he called "The Orion Project" with a spaceship to be powered by 2,000 atomic bombs to take him to Mars, Jupiter, and Pluto. Is it feasible?

Bethe: I think so. The objection would have been that nobody likes 2,000 atomic bombs exploded in the atmosphere.

Would you have gone on this spaceship?

Bethe: That's not the kind of thing I like to do. I don't think I would even want to go on the much safer shuttle.

THE SUN

Your 1938 theory that the beta carbon cycle required five million years to complete six linked transformations was the first explanation of solar and stellar energy that met all the known facts. Was Rutherford of any help?

Bethe: He had nothing to do with it whatsoever.

What about his fusion of helium and hydrogen idea?

Bethe: Why, you mean to make helium from hydrogen? That was obvious once you knew the atomic weights.

So Rutherford was no help.

Bethe: Well, nuclear physics was of help to me. Rutherford and hundreds of others taught us what a nucleus consists of and what it depends on and so on. The idea of hydrogen into helium was trivial. There's nothing to it. The question is how to do it.

Rather like Einstein's $E = mc^2$ says nothing about how to split the atom, am I right?

Bethe: Yes.

Was your discovery immediately accepted by your peers?

Bethe: Yes.

Can you describe for a non-scientist how you discovered what causes the energy in stars?

Bethe: To any physicist in the 1930s, that was obvious. The main thing is that in the helium nucleus, protons and neutrons are very tightly bound. So when you combine two protons and two neutrons to make a helium nucleus, you set free a lot of energy. Solar energy is caused by a combination of two protons, which make a deuteron, plus a free, positively charged electron, and the neutrino. The center of the sun is a plasma of photons and electrons at a temperature of 14 million degrees. The carbon cycle from a weak interaction produces energy in the stars that are hotter and more massive than the sun.

Are the source and makeup of cosmic rays a mystery?

Bethe: It is a mystery just how they are produced. Once they are here we know perfectly well in great detail their composition.

What was your contribution towards radar?

Bethe: I came to it quite late compared to Oliphant and Watson-Watt. Our problem in 1942 and 1943 was to refine the process. What I perhaps contributed to most importantly is how to measure the intensity of radar waves. The British were way ahead of us and they told us, "Now boys, make it better."

Were you in favor of SDI or Brilliant Pebbles? [Proposals to have a defense in space against nuclear attack]

Bethe: No. Because they're no good.

Is nuclear power a safe, permanent source for our energy needs, and are today's safeguards good enough?

Bethe: Yes.

Why haven't we used the sun more as an energy source?

Bethe: Because it's a very diffuse source. You need many square miles to replace a single power plant. In nuclear power you have a very concentrated source. Solar heating is fine for heating a house, because you don't need very much energy then.

In 1947 you wrote, "Neutrons and protons within an atomic nucleus are held together by the most powerful forces known. Remarkably little is understood of the nuclear forces, even after many years of research. We know they are enormously strong, more than a million times stronger than the chemical forces that hold oxygen and hydrogen together in the water molecule. While these chemical forces in turn are many times stronger than the elastic forces which hold together a lump of steel." Is it a great deal more understood today?

Bethe: Yes, thanks to many people. Among theoretical physicists, a very important one was Murray Gell-Mann of Caltech.

Would you like to see the supercollider built?

Bethe: Yes, because it would get us information at higher energies which we can't get any other way. We want to know are there more particles, and what particles, how are they made, and so on. So we hope to understand what holds the nucleus together in a more fundamental way. [The supercollider was voted down because of persuasive evidence that for various reasons the actual cost would greatly exceed the estimated cost.]

Is there much to discover?

Bethe: Yes. We have all these particles, but we don't know how they get their masses.

How many particles have we identified?

Bethe: Fundamentally, maybe twenty. [I mentioned this figure to Gell-Mann, who said, "He counted differently. There are about sixty."]

Do you approve of more atom bomb tests?

Bethe: A certain number would be good, to make existing bombs even safer. They are very safe already, but you will remember when the airplane [accidentally] dropped a few bombs into the sea. We want to be sure if such a thing happens again that not only is there no explosion but no spreading of plutonium. Plutonium is poisonous. So the two labs in Livermore and Los Alamos are engaged in designing tests directed toward making still more certain no unpleasant substances will be spread in an accident. I'd like those tests to be made. Although there is a Congressional resolution, the Hatfield amendment, which says all tests must stop in 1996.

When they're safe would you agree to a moratorium on testing?

Bethe: To a complete cessation of all nuclear tests. I want all countries to agree to stop testing completely from a certain date on.

HEISENBERG

What did you think of Powers' book, *Heisenberg's War*?

Bethe: He did a very good job in finding out how things happened. I disagree with Powers on the final conclusion. It has to be separated into two parts. 1. Did Heisenberg at any time want to make an atomic bomb or engage in research leading to that? The answer is no. On that I agree with Powers. The second point is, did he not engage in such research because he knew it was too difficult and would take too long, or did he do it entirely for moral reasons? And here Powers and I disagree. He says it was all for moral reasons. I think he said himself in the Farm Hall transcripts: it was because he was convinced it couldn't be done—a very different reason.

Doesn't Powers say Heisenberg deliberately sabotaged the effort?

Bethe: Yes, and I do not believe that.

Did you speak to Heisenberg about it after the war?

Bethe: I visited Heisenberg in 1948, and we talked about it then. The main thing he said was he wanted to save young physicists for the time after the war. And that I believe. But that's certainly only part of the story.

He also told Planck he wanted to preserve an "island of virtue" or something like that in Germany. (In fact, Planck persuaded Heisenberg to stay and try to form "islands of stability.")

Bethe: But an "island of virtue" doesn't do much good when there's a Nazi government.

When Einstein was asked after the war if he wanted to send any greetings to German scientists, he mentioned Von Laue and no one else. Would you have added any names?

Bethe: Certainly I would have greeted Heisenberg and Hahn.

How about Planck?

Bethe: At least a dozen others.

Weisskopf says his idea of kidnapping Heisenberg wasn't very serious, but he wrote a three-page letter to Oppenheimer suggesting the idea.

Bethe: We wanted to have it explored. At the time we proposed it, 1942, it would have made sense. It made no sense in 1944 (when the war was almost over).

When Moe Berg went over to kidnap or even kill him if neccessary. But didn't do either.

Bethe: We must be grateful to Mr. Berg for having sense.

You never feared you might be kidnapped by the Germans?

Bethe: I never thought so. At Los Alamos there was very little chance. I was a very less important person than Heisenberg.

Yet *Time* magazine called you one of Nazi Germany's greatest gifts to the United States.

Bethe (chuckled): That was very nice of them.

If Germany hadn't got rid of Jewish scientists, could they have made the atom bomb before the world's end?

Bethe: I don't think so. It was more an engineering and industrial job.

Did any of the Farm Hall transcripts surprise you?

Bethe: I was surprised how ignorant the Germans were. And that is very important vis-a-vis Powers. He thinks they knew a lot and even quotes me on that. And that is totally wrong. I think they were profoundly ignorant.

Were you surprised that Hahn was so upset his ideas led to the atom bomb that he broke down and cried?

Bethe: No, that didn't surprise me. Probably I would have cried, too.

Did Heisenberg say to you, "I don't blame German emigrés for working on the atom bomb in the U.S. Their hate for everything German was justified. They had to make some effort to prove their worth to their host countries."?

Bethe: I read that in the transcript. I don't remember this being said to me directly.

It's a quotation from David Irving's book, *The German Atom Bomb*.

Bethe: It's possible. More significant it's in the Farm Hall transcripts that Heisenberg did not blame the Allies nor the German refugees.

Wasn't it Hitler who made the German atom bomb unlikely because he believed in Blitzkreig, to get everything over quickly?

Bethe: Absolutely.

Which precluded long-term research on weapons.

Bethe: Right. But you see from the economic point of view, Germany could only win if she did it very quickly. So it was logical for the Germans to try that.

Goudsmit argued in 1947 that German scientists very much wanted to build an atom bomb, but had failed out of arrogance which blinded them to their own errors.

Bethe: I think there was an arrogance, but the main point was insufficient resources, insufficient industry.

Robert Jungk believed German scientists conspired to keep the atom bomb from Hitler.

Bethe: I don't believe that.

David Irving thinks the Germans concentrated on a nuclear reactor and simply never reached a point of discussion to make or not to make the atom bomb.

Bethe: That's correct.

Were you disappointed when Oppenheimer told the authorities in secret testimony that Bernard Peters (a friend) was dangerous?

Bethe: I wrote him a letter about it. I don't know what the merits of the case really were. Oppenheimer very much wanted to protect himself. I guess in some cases he was indiscreet about his friends.

I thought one of Oppenheimer's strongest ethical rules was not to harm anyone.

Bethe: Yes, that's true.

I'm quoting you now. Do you still believe Oppenheimer was "superior in judgment and superior in knowledge to all of us at Los Alamos."?

Bethe: Did I say that? Well, I think it's correct.

Some called him an authentic genius because he always gave you the right answer before you asked him the question.

Bethe: That's right. (Though when they were graduate students together in Germany, Bethe detested Oppenheimer for pointing out his slight math mistake at an open meeting.)

More of a genius and more knowledgable than Bohr?

Bethe: Oh, no. That's very different. But Bohr never tried to run a big project.

What do you think was Teller's motive for attacking Oppenheimer?

Bethe: You'll have to ask Teller.

Apparently Oppenheimer never praised Teller. If he had, someone suggests, the destiny of both men would have been different.

Bethe: It's a difficult business. Teller didn't feel sufficiently recognized at Los Alamos.

This is from Peter Wyden's book, *Day One*: "When Kitty, Oppie's wife was away and he called on the friends who were looking after his daughter, Toni, Robert didn't ask to see the baby and when she was brought out he never touched her. Eventually he wanted to offer Toni to friends for adoption, which appalled Kitty."[2] Do you believe this is true?

Bethe: I have no idea.

But you knew him very well didn't you?

Bethe: Yes. And when she was older he was devoted to her. I don't know if it's true, and if it is I can't explain it. (Toni Oppenheimer committed suicide at the age of 32 in 1977, reportedly over an unhappy love affair.)

SPECULATIONS AND ESTIMATIONS

What's your view of Superstrings?

Bethe: Maybe in the course of time it will get somewhere. At the moment it's a hope rather than a theory.

Do you believe in the Big Bang or in continuous creation?

Bethe: I believe in the Big Bang or the modification of it Alan Guth suggests.

Any hope for Einstein's unified field theory?

Bethe: I don't think it's completely hopeless. It certainly won't be found overnight. And Einstein was on the wrong track because he didn't use the quantum theory.

John Wheeler says there's still hope.

Bethe: Well I don't think he'll find it.

How about intelligent life on other planets?

Bethe: There must be lots of planets in the world and conditions on some must be favorable to life just as on earth. To what extent life develops toward intelligence is very hard to tell.

Did you change your mind about working on the H-bomb only because you believed the Soviets would soon make it, or because of the outbreak of the Korean War?

Bethe: It was the first reason. It was mainly the realization that with the Teller–Ulam discovery it was quite easy to make it, therefore the Russians could make it, too.

Victor Weisskopf told me that Fuchs inadvertently gave the Russians faulty information about how to make the H-bomb.

Bethe: That is probably true. But the important thing was he gave them very detailed and complete information on how to make an atomic bomb, except for the production of plutonium, which he gave them later when working on British atom research. In fact, the Russians have now published, this year (1993), the fact that their first atomic test in 1949 was a carbon copy of our test at Trinity in 1945.

What did you think of Fuchs when he was working for you?

Bethe: He was an extremely hard worker. He volunteered to do extra duties.

And your reaction when you heard of his spying?

Bethe: I was terribly shocked to learn of Fuchs' betrayal. I felt personally betrayed. On the other hand, he's the one scientist I know who changed history.

Who was the greatest scientist, Einstein, Bohr, or Heisenberg?

Bethe: Of this century, Einstein first, Bohr second, and Heisenberg third.

GERMANY

As a boy did you find, like Einstein, that German education was too much by rote?

Bethe: Yes, there was some of that. There is less in this country. On the other hand, we got a very good education, and when we left high school we had a good deal of knowledge and technique of working.

You left Germany in 1935 but visited every summer until 1938. Did you witness any overt anti-Semitism there?

Bethe: That's an interesting question. You mean that the Jews were beaten or something?

Or insulted. Or you were insulted.

Bethe: I myself was not insulted. But I lost two jobs. I knew many people who suffered greatly and that includes contemporaries. A young man I knew, Mark, was taken to a concentration camp. Three of my relatives were taken to concentration camps, internment camps, and two died. So I knew a great deal in 1936 when American visas were still available. My friend got an American visa and on that basis he could leave. They freed him from the camp because of the visa.

Do you know much about contemporary Germany and anti-Semitism?

Bethe: I know more from the newspapers than from experience. The papers had a lot of bad stories to tell. I think in some places and at certain times it's been very bad and is very bad. Fortunately, the government is now taking measures against it. But anti-Semitism persists.

Did any Germans apologize to you for anti-Semitism?

Bethe: After the war that was customary. In fact, we went over in 1948 and nearly every German we had known before the war, when we met again, apologized.

Can you explain anti-Semitism?

Bethe: It is so satisfying to have somebody to despise. The scapegoat.

Did any of your family remain in Germany during the war?

Bethe: Yes, indeed. My father [who was not Jewish] did, and had a second set of children, two of them. And I'm in contact with them.

They were not pro-Hitler, I presume.

Bethe: They were certainly not.

EVERYDAY LIFE

What papers and journals do you read? And what TV do you watch?

Bethe: One and only one newspaper. *The New York Times*. Occasionally I look at *The New Yorker*. I watch TV approximately once a year: the elections and the Winter Olympics. I watched MacNeill-Lehrer once or twice. There were very good things in it, but I also was bored so why do it? It makes no difference to me whether I know things 12 hours sooner.

What do your children do?

Bethe: My son is a banker. My daughter is as far away from me as possible. She lives in Japan. She's interested in Japanese Noh plays and is quite an expert. She's written books about those. I love her dearly. But their interests are certainly far away from mine.

Are they interested in your achievements?

Bethe: My son is interested in science, and that includes my work. My daughter is not.

You're from a family of academics. Whom do your children take after?

Bethe: I have a great uncle who was a banker. But really Henry [his son] chose that subject because he didn't want to go into physics. It would have been too competitive.

Do they have children?

Bethe: My son has a son who is a student at Exeter. He's an all round boy. He's quite good at academics, but he's more interested in football and such matters. My daughter has very young children. Her boy is 8 and her girl is 6, so I don't know what they'll be interested in.

Weisskopf says if he wasn't starting his career now he'd study neuroscience—the brain. How about you?

Bethe: I do not know. Certainly Weisskopf's choice is a very good one. Intriguing. Possibly I would go into biology.

Are you religious?

Bethe: I am an atheist.

How do you get your ideas?

Bethe: Mainly by myself. I do go for walks but mostly with my wife [Rose Ewald, daughter of a Nazi-exiled German theoretical physicist, Paul Ewald]. And walks are very good because they stimulate me generally. And lately I've found some answers during a walk. But mostly in my office.

Any big regrets about your work or life?

Bethe: Whatever went wrong, soon went right.

Chapter Six.   John Wheeler

Physics is "a magic window. It shows us the illusion that lies behind
reality and the reality behind illusion."
—John Wheeler

John Wheeler's parents were used to answering questions; they were librarians. But until he came along they had never faced non-stop interrogation. The young boy would ask: Why this? Why that? Does anything last for ever? If I keep going out in space, will I ever reach the end? Relief came when he learned to read. Then they could respond to his questions with books. At ten he got one that settled his future, J. Arthur Thomson's *Outline of Science*.

At the same age he found a dynamite cap near his home in Youngstown, Ohio. Curiosity overcoming caution, he picked it up and put a match to it. The explosion took off the tip of a finger. So began his lifelong preoccupation with explosions. No scientist has had more to do with the creation of the atom bomb and the hydrogen bomb or is more informed about the devastating explosions of dying stars that may herald—some 20 to 50 billion years ahead—the end of our universe.

A prize debater at Johns Hopkins where he got his Ph.D at 22, Wheeler was also, surprisingly, president of the Baltimore Federation of Church and Synagogue Youth. In 1934, he studied with Niels Bohr in Copenhagen. Four years later, having recently become assistant professor of physics at Princeton, he and Bohr published their joint effort, *The Mechanism of Nuclear Fission*. In it they explained how uranium-235 could be used in an atom bomb, or as fuel in a nuclear reactor, a couple of years before Heisenberg got the idea.

Modest about his own triumphs—groundbreaking work on nuclear structure, elementary particles, the theory of gravitation—Wheeler has compared Bohr not unfavorably with Buddha, Jesus, Moses, and Confucius.

His friends and relatives read like a roll call of great twentieth-century scientists. His teacher Niels Bohr, student Richard Feynman, and colleague Werner Heisenberg became his friends. A relative of Wheeler married Eugene Wigner, whose sister, in turn, married Paul Dirac. Among those five scientists Wheeler alone has not won the Nobel prize, though if a show of hands by his scientific peers would do the trick he'd have it. Heinz R. Pagels, Executive Director of the New York Academy of Sciences, has called Wheeler "that visionary of American theoretical physics."

At 77, after several years as head of the Center for Theoretical Physics at the University of Texas, Austin, Wheeler returned to Princeton. He brought firecrackers with him which he keeps on his desk ready to celebrate any discovery or appealingly audacious thought this side of mysticism. When he's away from his desk on working vacations in Maine and a bright idea surfaces, he fills empty beer cans with sand and fires them from an antique cannon into a bay near Boothbay Harbor.

His parents would recognize the curious three-year-old in the man, who says, "I am willing to go anywhere, talk to anybody, ask any questions that will make headway." He is equally open to answering questions.

This physicists' physicist asks: "Is man an unimportant bit of dust on an unimportant galaxy somewhere in the vastness of space?" And answers: "No. The necessity to produce life lies at the center of the universe's whole machinery and design." Yet he also speculates that the whole machinery and design is destined to be annihilated.

How does he choose what to study? "In my field," he said, "find the strangest thing, and explore it." Wheeler follows his own advice. Not much could be stranger than a black hole—he coined the term in 1938 and it caught on. *Black holes* are the graves of dead stars: cosmic traps in space from which neither light nor matter can escape. What happens in black holes is Wheeler's main concern although he has frequently been diverted by what happens on this planet. For example, in a meeting of an American-British committee during World War II to discuss security, Wheeler warned that a trusted individual motivated by a crazy ideology could sabotage the Manhattan Project. Listening attentively, sitting across from Wheeler, was a member of the British group. His name was Klaus Fuchs, and a few months later he was arrested for giving secret atom bomb information to the Russians. He confessed and spent nine years in prison, before resuming his career in Communist East Germany.

In 1979 Wheeler flew to West Berlin to celebrate Einstein's centennial and noticed with surprise that Fuchs was listed among those attending. "To have an adventure," Wheeler arranged to meet a man he regarded as the greatest spy of all time. But to avoid shaking hands with Fuchs, Wheeler held a notebook in one hand and a cup of coffee in the other.

I asked: Did you avoid shaking hands with Fuchs because he was a traitor?

Wheeler: I just didn't want to be involved in shaking hands with him.

What did you learn?

Wheeler: He simply told me about his work with the East German Atomic Energy Program on power plants.

Is there any equivalent today to the atmosphere of Niels Bohr's Institute in the 1920s and '30s?

Wheeler: Little groups here and there, but none that has the radiation of his personality and the group that he had circulating.

Wolfgang Pauli called Bohr an idiot and told him to shut up. Was that your experience of how things went?

Wheeler: Pauli had gotten rather better adjusted to life when I knew him because he had married a solid, wise woman. Of course, everybody at the Institute was open to argue absolutely.

With the exception of Richard Feynman, I can't imagine a student at Texas or Princeton calling you an idiot.

Wheeler: I have had two students who thought I was cuckoo. In fact, I glued a statement of one into one of my notebooks as a salutary warning to be aware of my shortcomings and failures. I thought it was healthy.

After a discussion with Bohr far into the night, Heisenberg thought, "Can nature be as absurd as it seems to us in these atomic experiments?" Do you think nature is absurd?

Wheeler: No, I see it as exciting. The way I put it is: The quantum is the crack in the armor that covers the secret of existence.

Did you have similar discussions with Heisenberg?

Wheeler: The first time I met Heisenberg was in the garden in front of the Bohr Institute between the house-like building and the street. He was very open and friendly. I was then 23, so I was the oldest person at the Institute [except for Bohr]. Some years later, Heisenberg and his wife stayed at my house in Princeton. We were discussing old times and great questions, and he told me that after he had written his paper on the uncertainty principle it made quite an impact around the world. I believe its publication was before Bohr himself had given his September 1927 address on complementarity. In the interval between the spring and fall of the year, Heisenberg was out sailing with Niels Bohr and Nields Bjerrum, the great chemist. And Bjerrum asked Heisenberg to tell him about the Uncertainty Principle. After Heisenberg had finished, Bjerrum turned to Bohr and said, "But Niels, that's what you've been telling me ever since you were a boy."

What do you recall most vividly about Bohr?

Wheeler: He needed to have somebody to walk up and down with him talking intensely and giving him his full attention. As a matter of fact, last week when I was in Bern, Switzerland, I met my old friend Rene Mercier. He had been at Copenhagen in 1932-3, a year before I was there, and had married a Danish girl. Mercier spoke to me of the Pompeian court at Carlsburg. It's colonnaded, part of the House of Honor, and the whole thing is indoors under a glass roof: the idea of J. C. Jacobsen, the great man and founder of the Carlsburg Breweries.

It's the first steel-strutted building of that kind.* Walking around with Bohr two days before the outbreak of World War II, at a time when it was clear war was inevitable, Mercier said he noticed tears streaming down Bohr's face because he knew what was coming.

Heisenberg was also with Bohr in Copenhagen. Was he as great a scientist as Bohr and Einstein?

Wheeler: A 1985 book, *The Foundation of Physics*, has three names on the dedication page: Bohr, Einstein, and Heisenberg. Heisenberg remarked that he would never have been able to discover the matrix formulation of quantum theory—which he did—without the mathematics he got from Gottingen; the idea that you can indeed make progress in physics he got from Sommerfeld, the physical insight he got from Niels Bohr. But in that connection he did not mention Einstein. Lev Landau had a categorization of physicists [Landau won the 1962 Nobel Prize for pioneering theories about liquid helium]. My memory is that he had Einstein and Bohr at the top and nobody else there. I'd go along with Landau and put no one on the same level of the totem pole as Einstein and Bohr. The debate between them, to my mind, is the greatest debate in intellectual history. In thirty years I've never heard of a debate betwen two greater men over a longer period of time, with deeper consequences for understanding this strange world of ours.

You once said: "We go down and down from crystal to molecule, from molecule to atom, from atom to nucleus, from nucleus to particle, and there's still something beyond both geometry and particle. In the end we come back to mind. How can consciousness understand consciousness?" This seemed to you a paradox. But you also quoted Bohr as saying we only make progress from paradoxes. What did he mean?

Wheeler: That we find out where we are wrong.

Roger Sperry, an expert on the brain, gave up experiments on the brain to speculate on the role of the mind.

Wheeler: Some day I hope to learn what the difference is between the mind and the brain. I've never understood what the purported dif-

* Jacobsen had the building constructed in the Pompeian style after visiting Pompeii. Bohr was the second man invited to live in the House of Honor, as Denmark's greatest citizen.

ference is. We humans have limitless power to confuse ourselves with words.

What do you think of E.H. Walker's speculation that our universe is inhabited by an almost unlimited number of rather discrete, conscious, usually nonthinking entities that are responsible for the detailed workings of the universe?

Wheeler: That sounds like demonology.

Do you share George Wald's view that consciousness suffuses the universe?

Wheeler: I'd like to see the evidence. You can always define consciousness in such a way that that would work.

Paul Dirac told me he was working on a theory that as the earth ages, so gravitation weakens. Was he on the right track?

Wheeler: There's enormous work on it. Among others, Freeman Dyson has contributed to that effort. But the evidence shows with ever greater precision total lack of any such change. I had useful discussions with Dirac. He was not one to expand the realm. As somebody once said: "He shines a bright light within a circle, but there's no penumbra around it."

Was Richard Feynman your most brilliant student?

Wheeler: Brilliant isn't the word to use if you mean brilliant to be synonymous with creative. Bohr [came up with a few] words which I translated into English and they apply to Feynman:

"If? Who hasn't?
Talent? Toy for children! Commitment makes the man; Only diligence, genius." And Feynman had a marvelous combination of commitment and diligence as well as insight.

This is like 99 percent perspiration, one percent inspiration.

Wheeler: Right.

What's your most vivid memory of Einstein?

Wheeler: A call I made on him in 1941 to explain "the sum of histories" approach to quantum mechanics being developed by Richard Feynman, then a graduate student at Princeton. I had gone to see

Einstein hoping to persuade him of the naturalness of the quantum theory when seen in this light, connected so closely and so beautifully with the variation principle of classical mechanics. He listened to me for about 20 minutes until I had finished. Then he repeated his now familiar remark, "I still cannot believe that the good Lord plays dice." And then he added in his beautifully slow, clear, well-modulated and humorous way, "Of course I may be wrong, but perhaps I have earned the right to make mistakes."

Have any recent discoveries made Einstein's unified field theory more likely?

Wheeler: It's come to life in an absolutely spectacular form in the last decade [1980s]: the so-called superstring theory. It's the idea that in addition to the dimensions you and I see and work in, three in space and one in time, there are additional dimensions the geometry of which, however, is curled up into a very tight sphere, everywhere and at every point. Each little sphere is in a certain sense like an organ pipe. Each organ pipe has air vibrating within it which gives it musical tone, depending on its length and diameter. Each little sphere in the geometry has its own vibrations, more numerous than an organ pipe. And these vibrations describe the various fields of force that give rise to the particles we know. Period. The theory is a result of a succession of ideas that began with an associate of Einstein, Theodor Kaluza, and an associate of Niels Bohr, Oscar Klein—the so-called Kaluza–Klein, a five-dimensional theory in 1926. Nowadays the idea is that there are additional dimensions. There's some question whether one should have four plus six dimensions, or four plus seven dimensions. At the moment the favored dimensionality is four plus six. I think Einstein would have found the theory very appealing. The Kaluza–Klein idea has been generalized in a perfectly wonderful way today. So that at any rate Einstein's dream, his hope, his goal of reducing the various forces of physics to pure geometry has taken on life. It's unbelievably active today. I don't know any topic in theoretical physics today that's more investigated by more able people.

Through this ten-dimensional theory would Einstein have been able to reconcile himself to quantum mechanics?

Wheeler: No, but he would have to take a new look at the quantum theory.

Except to teach us more about the universe, the 10-dimensional theory doesn't have any effect on our daily living does it?

Wheeler: Some day what we learn will, I feel, be the most revolutionary thing in its effect on human outlook.

Wasn't Einstein's final opinion that although he couldn't accept quantum theory intuitively, rationally he admitted it was probably right?

Wheeler: It's a very ticklish job to say what he thought with exactly the right semantic tone to it. But that's roughly it.

Do you still believe the "many worlds" interpretation of quantum mechanics?

Wheeler: I have written, "Why not?"

Does it supersede or complement the ten-dimensional world theory?

Wheeler: The ten-dimensional theory applies to the micro world. Initially I supported it, but I have changed my view, giving various reasons, but especially too much metaphysical baggage.

How do you define physics?

Wheeler: As a magic window. It shows us illusion that lies behind reality and the reality behind illusion.

By reality behind illusion do you mean the dancing atoms, for example, behind what appears to be a solid table?

Wheeler: Absolutely.

Then what is the illusion behind reality?

Wheeler: That would be the elementary quantum phenomenon, when a photon goes from a slit in an interference device and is registered on a photographic plate. [Known as the double-split experiment. In his lectures Feynman described it as "a phenomenon which is *absolutely* impossible to explain in any classical way. and which has in it the heart of quantum mechanics. In reality, it contains the *only* mystery . . . the basic peculiarities of quantum mechanics."

Light shone through a vertical slit in a screen appears on the wall behind it as an illumination resembling the full moon. When shone through two adjoining slits in the screen it should appear on the wall behind as the sum of the light from the two slits. Instead it appears as alternating bands of light and darkness known as interference pattern. This indicates that light is wave-like because only waves can create interference patterns. But Einstein had shown through the photoelectric effect that light consists of particles. How could light be both waves and particles? An even bigger mystery is how light (photon or electron), in another experiment, appears to "know" where to go—almost as if it were conscious! The wave-particle duality defied the laws of classical causality and led the way to quantum theory.] The illusion is the idea that the electron has an identifiable path. The reality is the point of entry and the moment of registration. There is a wonderful thing when this double-slit experiment was the center of discussion between Heisenberg and Bohr in Copenhagen. This was before Bohr occupied the House of Honor. The first occupant was the great Danish philosopher, Harald Hoffding, and Bohr's favorite professor at the university. Hoffding invited Bohr around to discuss this double-slit experiment and he asked Bohr, "Where can the electron be said to be?" [In its passage from the point of entry until the point of registration]. Bohr's answer was: "To be? What does it mean, to be?" I don't know of any more dramatic way to say it than that.

In other words it's pretty enigmatic. It's using a language that's not appropriate to the phenomenon?

Wheeler: The words "to be" are tricky words when you get down to it in nuclear physics.

Going from micro to macro, what's your view of the universe?

Wheeler: The universe does not exist "out there," independent of us. We are inescapably involved in bringing about what appears to be happening. We are not only observers; we are participators. In some strange sense this is a participatory universe. Physics is no longer satisfied with insights only into particles, fields of force, into geometry, or even into space and time. Today we demand of physics some understanding of existence itself.

You and Bohr jointly published a paper on the mechanism of nuclear fission that helped lay the groundwork for various uses of atomic energy. Your student, Richard Feynman, admitted that he'd gotten a main feature of his winning idea that gained him the Nobel prize— considering that there's only one electron in the universe—from you during a phone call. In fact, he said so, after accepting the prize. Did he ever offer to name you as coauthor of the work? If not, should he have done?

Wheeler: Feynman never wrote my suggestion down as such. The work he did later went beyond my idea.

Feynman said of you, "His ideas are strange: I don't believe them all. But it is surprising how often we realize later he was right." Was the electron one of those strange ideas?

Wheeler: One that was right? Right.

In 1946 you established the hypothesis of the positronium, a short-lived atom consisting of an electron and positron circling about a common center. Is it confirmed?

Wheeler: Yes, and in addition a molecule made of two electrons and one positron, confirmed by Alan Mills at Bell Labs.

Project Matterhorn, a top-secret enterprise in 1952 was to develop weapons based on thermonuclear fuel. Edward Teller credited your drive and optimism as important factors in its success. Is it still top secret?

The constituents of the hydrogen bomb are still secret. The existence of one is not, of course.

How about the gravitational-electromagnetic entity, a ball of light radiation held together by its own gravity and called "the geon"? Did your work track it down?

Wheeler: It predicted that such objects in principle can be made. But nobody has yet made one. Point one. And point two: that such an object is intrinsically unstable in the sense that it will either blow up, explode outward, or collapse inward to make a black hole. It's simply an example of mass without mass; that's why I find it so interesting.

Is a neutron also an example of mass without mass?

Wheeler: We don't know the whole story about the internal constitution of the neutron. We have quarks inside, but we think of quarks as having mass. But when you get down to the bottom of things, again Bohr's question about what is real is useful: "To be? To be? What does it mean, to be?"

You said the "most worrisome single problem in theoretical physics today is the phenomenon of gravitational collapse—black holes." You coined that term, didn't you?

Wheeler: Yes. Of course there has been a Black Hole of Calcutta before that. The first to suggest the existence of black holes was the Reverend John Michell of Queens College, Cambridge, in 1783.

How did you get from your more practical work on the hydrogen bomb to investigating the nature of black holes?

Wheeler: Through his work in quantum electrodynamics, Richard Feynman solved the problem of how charged particles interact even across total vacuums and unlimited distances. I then studied how uncharged electrically neutral bodies such as stars and planets interact over great distances. That led me to the possibility of universal gravitational collapse. A brief expression for that is crunch.

The total destruction of the universe? How soon?

Wheeler: Not for another 20 to 50 billion years.

What do you expect to happen?

Wheeler: The expansion of the universe [following the Big Bang] will be arrested by gravitational forces, galaxies will reverse direction reaching almost the speed of light until the universe has reached a state of enormous compression.

You believe a similar thing has occurred in black holes—after the death of stars, enormous compression. Stephen Hawking calls you the most articulate proponent of black holes. Could you give a brief account of them?

Wheeler: Hydrogen atoms are brought together in space under great pressure through gravitational attraction. As a result great heat is generated, thermonuclear burning occurs and then a star is born or "turns on." During this hydrogen burning hydrogen is converted to helium, helium to carbon, and carbon into heavier elements.

Then, as you say, "The still mysterious process we call life occurred when the atoms from the stars were rearranged into molecules, cells, proteins, etc." Will this life continue for at least another 20 billions years before the crunch, and barring any other major catastrophe?

Wheeler: Right.

Are black holes your specialty?

Wheeler: My specialty is everything.

Stephen Hawking shares your interest in black holes.

Wheeler: Yes, we are interested in the same subjects. We have talked about various things. But I generally have to talk with somebody else as intermediary because I have such trouble understanding what he says. Now he uses a speech synthesizer and his assistant interprets from that. [Stephen Hawking, regarded by some as the most brilliant theoretical physicist since Einstein, holds the chair at Cambridge once held by Newton and Dirac. He suffers from Lou Gehrig's disease which confines him to a wheelchair. After a tracheotomy during a bout of pneumonia in 1985 he lost his voice.] I had to preside when he gave a lecture in Chicago a few years ago, so I could look over his shoulder as he was using his voice synthesizer. He had composed the bulk of the talk beforehand, but he brought the gadget along for the questions.

Murray Gell-Mann won the 1969 Nobel prize in Physics for his work on "quarks." He suggests they are the smallest particles in the universe. Do you agree?

Wheeler: The stepping stones to the ultimate constituents. But by themselves, ultimately, no.

You envisioned the ultimate indivisible particle as being much smaller than a quark. As I recall it was one millionth of a billionth of a billionth of a centimeter. Am I right?

Wheeler: Let me translate that into English. A billionth is ten to the ninth, and two of them is ten to the eighteenth. A million is ten to the sixth. Ten to the eighteenth . . . then ten to the twenty-fourth. Of a centimeter, did I say? No, it's smaller than that.

Should another billionth come into the equation?

Wheeler: Let's see. Three nines makes twenty-seven, don't they?

Yes.

Wheeler: Then add six to that and you get thirty-three. And that's what you want. So it's three billionths you need in your statement.

How did you come up with that figure?

Wheeler: By looking at the story of quantum fluctuations in the geometry of space. To predict and understand the electromagnetic fluctuations were the greatest achievements in the postwar [World War II] era. To go from the fluctuations in the electromagnetic field to the fluctuations in space geometry itself was what led to the quark length.

But how can the layman understand something which has no before or after?

Wheeler: It is that a concept of time loses its meaning there.

I see.

Wheeler: That's great. I'm glad you do.

No, I say "I see", meaning I heard what you said. But I still don't know what it means.

Wheeler: Right. Well, then, we're even. [He laughed.]

Still talking of enigmas, have you read *The Dancing Wu Li Masters* by Gary Zukav?[1]

Wheeler: I didn't, because if I did I'd have to comment on it and whatever I said would be held against me. [Among other things the book shows the growing similarity between the language used by Eastern mystics and Western physicists.]

There was a *Reader's Digest* article about you in the September 1986 issue. It reported that Einstein and others noted in 1935 that pairs of particles originating from the same source but widely separated seemed to match or influence each other, because measuring one determined the distant properties of the distant other. In 1965, John S. Bell proved that no local or hidden variable could explain these correlations. And in an interview in *Omni* in 1983, Nobel laureate Brian Josephson on the same subject says: "In essence, the particles would be communicating instantly, faster than the speed of light . . . John Bell and, later, Henry Stapp used the well-accepted equation of quantum mechanics to show that such 'superluminal' communication is just what one might expect. The theorem raises the possibility that one part of the universe may have knowledge of another part—some kind of contact at a distance under certain conditions." In other words, says the *Omni* interviewer, "when a polar bear jumps into Arctic water, in some weird way it may cause a train wreck in the south of France." And Josephson did not disagree.

Wheeler: I'm afraid the wording confuses the matter. I don't agree that particles could be communicating instantly. I think we have the proper theory and we don't have to invent a new one. We have it already. The quantum theory describes things correctly, accurately, completely. I think John Bell would be the first to agree that his theory deals with worlds that never were and never can be.

Bell is in Switzerland, isn't he?

Wheeler: Yes. I had dinner with him there Monday night, last week. It's amusing that Murray Gell-Mann in his lecture recently at the Wigner birthday celebration in Maryland, spoke of the Bell business and the non-locality. And he said that anybody who thinks this non-locality is a real, physical non-locality in a real sense, is mistaken. It's like that photon or electron that Bohr and Hoffding were talking about. It didn't have a location.

In his 1988 book *Infinite in All Directions*, Freeman Dyson wrote: "I hope that the notion of a final statement of the laws of physics will prove illusory as the notion of a formal decision process for all mathematics. If it should turn out that the whole of physical reality can be described by a finite set of equations, I would be disappointed. I would feel that the Creator had been uncharacteristically lacking in imagination."[2] Stephen Hawking believes physicists are nearing the end of their quest for the fundamental laws of the universe. Do you agree?

Wheeler: No. Maybe we're 20 percent of the way.

Is electricity still a mystery?

Wheeler: It depends what you mean. I've just received in the mail a new textbook on electricity and magnetism, and it's loaded with examples and problems and methods of calculation. It would be hard to name any subject known to man about which we know more and yet also one about which we know less.

So it still is a mystery.

Wheeler: Everything is a mystery.

How about dreams? Do they interest you?

Wheeler: If my dreams were interesting they might. [He laughed.] I won't say they're pedestrian, but I've never solved a problem in a dream like Kekule.

How about hypnosis?

Wheeler: I've never tried it. I was present when Feynman let himself be hypnotized at Princeton.

British physicist Brian Josephson who meditates every day, believes there may be a universal intelligence which one can tune into. Others refer to this as God. I know of one scientist who said that the more he studied the more he believed in God, and of others who say the more they studied the less they believed in God. What do you say?

Wheeler: I'll translate it from the Latin: Concerning taste there can be no disputing.

Do you believe in a personal God?

Wheeler: The idea is a little too concrete for me. I think of divinity as being present everywhere.

You don't think of Christ as God?

Wheeler: Instead of running him down, I'd run others up.

Do you then, like Feynman and Pauling, call yourself an atheist?

Wheeler: I think nature is the safest guide to these questions, and in time the truth will be revealed to us.

What do you think of the possibility of life after death?

Wheeler: In my view there's one form of life after death that no one can deny; that each one of us lives again in life made good by his or her presence.

How about people like Hitler who made it worse?

Wheeler: Unfortunately, they, too, live again. That's the negative side of my belief. I also believe that no picture is a picture that does not have a frame, and the further idea that life without death would lose its meaning.

What's your response to people who claim to have had occult experiences?

Wheeler: They make me sad and thoughtful. One student told me that he frequently spoke with people who came down from Mars, and he invited me to join him on Route 206, ten miles north of Princeton, to meet the Martians. I wasn't curious enough to see for myself. But I have wondered what spurs people to take such a serious interest in Martians, flying saucers, the Bermuda Triangle, and similar fantasies. I was specially interested in the psychology of the investigator. "What is the source of the element of good judgment?" I asked myself. "What distinguishes real science from pseudoscience? And in studying these fascinating byways of self-delusion, where did one have to draw the lines and say, 'I'm not going to waste any more time on this?' "

Yet discoveries in real science seem to be getting more and more fantastic. Don't you think physics has become more exciting than ever before?

Wheeler: Oh, absolutely, absolutely.

Roger Sperry says, "Most scientists as they get older and see the end approaching no longer have the patience to waste their time on the kinds of things they formerly thought they could do for ever. They raise their perspectives with age." Agree?

Wheeler: It would be hard to escape that.

What are you working on now?

Wheeler: The thing I'm trying to find out is this: How is the world put together? If I were the Lord starting up the universe, what would convince me that I could make not a go of it without the quantum principle? What is the utterly simple, beautiful, compelling principle that underlies the quantum principle? That's the question.

It sounds like the sort of project Einstein would suggest.

Wheeler: Well, I think we have to face it.

In 1981 Stephen Hawking attended a conference on cosmology organized by Jesuits at the Vatican. The Pope told the scientists not to inquire into the Big Bang itself, as it was the moment of creation and therefore an act of God. Ironically, Hawking had just given a talk in which he had suggested that there was no moment of creation. Do you agree with the Pope's (belief) that the Big Bang is off limits, so to speak?

Wheeler: I don't agree with the Pope. Inquiring is what we are here for.

As you know, in 1928 scientist J.B.S. Haldane said, "Now my suspicion is that the universe is not only queerer than we suppose, but queerer than we can suppose." How do you, almost seventy years later, see the universe?

Wheeler: I think the secret will be so beautiful, so compelling, that we will all say to each other, "Oh, how could it have been otherwise? And how could we have been so stupid for so long?"

Many of your scientific peers think you deserve the Nobel prize.

Wheeler: The greatest prize would be if I discovered the secret of how the world is put together.

How long do you think it will take to find out?

Wheeler: It might be one year. Or it might be a thousand years.

Chapter Seven.   George Wald

"We already have immortality, but in the wrong place. We have it in the
germ plasma; we want it in the soma (body)."
—George Wald
"We live in a world of chance, yet not of accident. God gambles,
but he does not cheat."
—George Wald

Einstein said of George Wald, "He has a strong mind." At
Harvard he was celebrated as a professor who taught biology with
joy and clarity, riveting his audiences with such remarks as, "Humans
are more like yeast than unlike it, because yeast and man have a common ancestor. Some of the ancestor's progeny became yeasts and
some became men, and those two journeys resulted in a change of
only 43 nucleotides out of 312."

The son of Jewish immigrants—his father was a garment factory
tailor—Wald's lifelong study has been vision. He was 26 in 1932,
when he discovered vitamin A in the eye's retina. Later he proved
that for humans to see in color they need three retinal pigments; all
made by a combination of vitamin A and one of three different proteins. In the 1960s he found that color blindness was due to a lack of
one of the three pigments. And in 1967 George Wald was awarded
the Nobel prize for establishing the primary physiological and chemical visual processes in the eye.

His passionate interests are far-ranging, embracing the human condition, human destiny, and the nature of our mysterious universe. An active member of the Union of Concerned Scientists, on a trip to the Soviet Union he discussed nuclear disarmament with then Soviet leader, Mikhail Gorbachev. In a *Washington Post* article, September 25, 1987, Dr. Wald deplored "some of the most extreme human rights violations in modern history by the government of the Iranian president" and called for sanctions against Iran to persuade the Iranians to end their war with Iraq.

During my conversations with him George Wald gave me his impressions of Einstein, Bohr, Oppenheimer, and Teller; his views on the universe, death, life after death, ghosts, gravitation and galaxies; as well as a subject that particularly engrosses him: "Life and mind in the universe."

ENCOUNTERING EINSTEIN

Wald: Soon after the end of World War II I was going from Harvard to a meeting in Princeton. I had a gap of a little time in the course of the meeting and hoped to see Einstein then. Philipp Frank, a philosopher of physics at Harvard knew Einstein well, and he wrote me an introductory letter to him. I very much looked forward to meeting Einstein. It wasn't just curiosity. You see, there was a *monumental* generation of physicists in the first half of the century. Their like doesn't exist now. I don't know why. I don't think it's because the protoplasm of physicists has changed since then. They were in constant correspondence, conversation, contacts of every kind. They climbed mountains together and cooked their meals over an open fire together. There was a continuous rubbing on each other. In the course of that each of them became something more than he would otherwise have been. And they felt they had the universe by the short hairs. Of that generation I knew only Einstein and Bohr.

On that first visit with Einstein I asked him would he please explain to me his friendly controversy with Niels Bohr involving the real meaning of the uncertainty principle? [In his uncertainty principle Werner Heisenberg showed it was impossible to determine simultaneously the exact position and motion of a subatomic particle.

This does not apply to larger material—from a grain of sand to a planet.]

And Einstein, very nicely and in a clear, elementary way, went ahead and did exactly that. This was our first conversation. He was a very easy person to converse with and rather jolly. He would say something and then lean back and laugh. He pointed out, how can one be sure, if one closes one's eyes, that the chair one has seen a moment before, on opening one's eyes one sees the same chair? And, though one believes that it must have been there all the time, that can't be demonstrated. Another time he said to me, "Science has become a Tower of Babel. All my life I have avoided this Tower of Babel." Soon after I was interviewed on TV and mentioned the Tower of Babel. One of my colleagues at Harvard said, "I heard you on TV last night. What was that funny thing you said about some kind of tower?" Einstein—and I'm one, too—was a Bible reader, and that's a great thing. I don't think he read it religiously. Certainly I do not. But it is very much at the base of our culture, and one can save a lot of time and space by referring to the Tower of Babel, or any of those bits of biblical mythology provided you are talking to people who are familiar with those myths.

What did Einstein imply by his Tower of Babel comment?

Wald: He meant that we are living in the midst of an information explosion. It began seriously after World War II. C.P. Snow helped it out. In his book *The Two Cultures*[1] he reproached the western world for not producing the scientists and engineers on the same scale as the Soviet Union did. After Sputnik, the west began to produce scientists and other technical people at an enormously increased rate. We're paying a heavy price for that. One finds very few—and I think this is what Einstein was referring to—broadly based scientists who can think with a wide view outside their own speciality. And the scientific literature has become completely unmanageable. Up until the end of World War II I had no difficulty in keeping up with physiology and biochemistry, which are my fields, and with biophysics. If anything really interesting happened in other branches of science it was easy to keep in contact. Now its growing increasingly difficult. I find that students of science and young scientists, unable to keep up with the literature, in desperation keep narrowing their fields. They become more and more specialized.

I met Einstein again in 1952 when I was giving the Vanuxem Lectures at Princeton, to which he came. Before the first lecture we were walking up and down the street in front of the lecture hall. Suddenly he turned and said, "For many years I wondered why the electron came out negative. Negative, positive—those are perfectly symmetrical concepts in physics. [The electron, discovered by J.J. Thomson in 1897 is a fundamental particle of electricity and matter, and electrons exist in all atoms as planetary particles revolving around the nucleus. The electron always carries a negative charge.] "So why is it negative?" Einstein said. "I thought about this for a long time. Finally, all I could think was . . . it won in the fight!" And I promptly said, "That's *exactly* what I think about those left-handed amino acids. They won in the fight!"

The fight Einstein was talking about was the conflict between matter and antimatter, between the negative electrons and the positive electrons, which had already been discovered and which in contact mutually annihilated each other. [The positron is a fundamental particle equal in mass and energy to an electron, but having a positive charge.] When an electron comes into contact with a positron there is mutual annihilation. The masses of both are annihilated and turned into radiation, according to Einstein's famous formula $E = mc^2$, in which E is the energy of the radiation, m the mass that has been annihilated, and c^2 this tremendous number. c is the speed of light, three times ten to the tenth centimeters per second. So, annihilating even a tiny bit of mass yields a lot of radiation.

As Einstein said in our conversation, positive and negative are perfectly symmetrical concepts in physics. So it is expected that exactly equal amounts of particles and antiparticles entered into the Big Bang. If they then were mutually annihilated we could have been left with a universe containing only radiation. No matter. How come we have a universe of matter?

You mean that according to Einstein's theory of mutual annihilation our universe should not exist?

Wald: That's right. So there are two possible solutions. One is that some of the astronomical bodies we see at really great distances might be made of antimatter. Everything close by seems to be matter, and not antimatter. But we can't be sure that astronomical bodies at great distances from us are made of antimatter rather than matter, because they would look to us exactly alike. All our information concerning them comes to us through radiation, and radiation doesn't care. Radiation consists of photons which, as we say, are their own antiparticles. So that's one thought: that perhaps somehow or other there are places in the universe which are made of antimatter, and other places like our own part of the universe which are made of matter.

But there is another much more engaging thought now. Two scientists at the Bell Laboratories, Arno Penzias and Robert Wilson, about 25 years ago discovered a very cold microwave radiation which fills the universe perfectly evenly.* It's apparently a very ancient radiation that has survived, and it goes equally in all directions.

The latest thought is that in the Big Bang that started our universe, what came into being were almost equal amounts of matter and antimatter, or particles and their antiparticles. In the fireball of the Big Bang packed to an almost unimaginable degree, these were in contact and there must have been an enormous fire-storm of mutual annihilation.

The present thought is that the radiation discovered by Penzias and Wilson is the residue of a fire-storm that came out of the annihilation at the time of the Big Bang. It's by far the most radiation that exists in our universe and, in fact, there are roughly one billion times as many photons of that radiation as there are protons and neutrons—the massy particles of our universe.

The present nicest thought is that what got into the Big Bang involved a tiny error of symmetry. And that error is that to every billion particles of antimatter there were one billion and one particles of matter. When all the annihilation was complete, one billionth remained. And that constitutes all the matter in our universe: all the galaxies, stars, planets, and all life. It's a strange and wonderful thought. And it would have been a result of Einstein's "fight."

In another conversation I had with Einstein at Princeton he was very sadly saying that his great hope was to achieve a unified field theory and he clearly was not going to succeed, and someone else would have to do that.

* See conversation with Arno Penzias in Chapter Eight.

One of my last meetings with Einstein toward the end of his life he looked at me with his face long and sad and said, "People keep writing to ask me what is the meaning of life? What am I to tell them?" That, like many things he said struck me very deeply. I've thought about it for many years. Finally I felt that I'd found a kind of answer. It is that the meaning of life is to keep asking questions like that.

[Philosopher Karl Popper called such ultimate questions as "Why was the world created?" and "Why are we here?" unanswerable. So did Samuel Johnson. Immanuel Kant, however, was less pessimistic, asking: "Why should nature have visited our reason with the restless endeavor whereby it is ever searching for answers, as if this were one of its most important concerns?"]

Can you compare Einstein with Bohr?

Wald: They were without question the greatest persons I have ever met. There was a little the feeling in the case of both of them of meeting an Old Testament prophet. They were at once the greatest and the most *childlike* persons I have ever met. They both had accomplished that wonderful thing of becoming wiser and more learned children as they grew older. One aspect of that was there were no fences around them, no boundaries beyond which they wouldn't go. They were interested in everything interesting. I thought sometimes of a man walking a puppy. The man walks a straight line, but the puppy's into everything. And they both went like the puppy. Bohr was a wonderful person who got fantastically excited by anything really interesting. I'm sure he and Einstein got along wonderfully well together. Both had a large capacity for enjoyment.

Abraham Pais, who knew them both, too, believed Bohr was a greater human being than Einstein. And C.P. Snow wrote, "Bohr was not as Einstein was, impersonally kind to the human race. He was simply and genuinely kind. It sound insipid, but in addition to his wisdom, [Bohr] had such sweetness. He was a loving and beloved husband and father."

Wald: I don't know much about their personal lives, but in my contact with Einstein he was an altogether sweet and concerned person, a lover of humanity. I would not make that kind of distinction between them. C.P. Snow I dislike in an impersonal way. He was obsessed with power, the whole problem of power and powerful people, and competition. His books deal in the most disagreeable people, painful people, all super-competitive, striving for power.

Your Harvard colleague, I. Bernard Cohen, suggested that Bohr's discoveries were more profound than Einstein's.

Wald: I wouldn't go along with that. They were the leading spirits in the reworking of what physics makes of the universe, and beyond that changed ordinary people's attitudes toward reality to a fantastic degree. I would say, if anything, Einstein had more effect on lay attitudes; relativity in some sense got to people as quantum mechanics could not. All of us live with space and time and gravity; the world of elementary particles and atomic structure remains hidden. Quite apart from specific, very important and basic contributions to physics, Bohr's principal cultural impact, perhaps—and I think he thought so—was in the complementarity idea, in the insistence that Heisenberg's uncertainty principle represents a technical failure but essential reality: that's the way the world is. I'd put the essence of the Bohr point of view about the uncertainty principle, which Einstein never accepted, in these words: It's not that we can't simultaneously specify the position and motion of an electron, but that it *does not have* a simultaneous specific position and motion.

Bohr extended the complementarity principle [which recognizes that to describe adequately the physical world, different, even mutually exclusive accounts may be necessary, such as light being both wave and particle] in the way Darwin extended his principle of natural selection to all human societies and human aesthetics—to a wide variety of social and intellectual experiences. I think he wanted it to have a broad cultural effect quite outside physics.

What great scientists have you known other than Bohr and Einstein?

Wald: Unknown to me until it was all over, during the months I spent at the Institute for Advanced Study at Princeton, Oppenheimer was being tried by the Gray Committee. [After being director of the atomic-energy lab at Los Alamos where the atom bomb was built during World War II, J. Robert Oppenheimer advised the government on atomic research until 1952 and became director of the Institute for Advanced Study at Princeton. In 1954, after a four-week security hearing during which fellow physicist Edward Teller equivocated, Oppenheimer was judged a security risk and stripped of his access to secret information.] So it was a difficult time. Oppenheimer was one of the most arrogant persons I've ever met. I say these things because I'm a scientist and the ins and outs of the souls of scientists have always fascinated me. Oppenheimer was a very vain person and with his vanity, arrogant. I remember getting the impression rather strongly that if he came into some formal conversation and someone else presented, before he spoke, some point of view which *he* would otherwise have expressed, he'd immediately change [Wald chuckled.] his view, and oppose them.

Still, almost all physicists came to Oppenheimer's defense, supporting him and not Teller.

Wald: Teller is really astonishing . . . I have read comments that his fellow physicists regard Teller with contempt, and that stemmed originally out of the Oppenheimer affair. Physicists were fond of Oppenheimer, and that speaks well for him. I had an odd experience sitting all together at a table, perhaps there were six people. Bohr was at the end of the table and Oppenheimer was sitting on his left. And that was amazing. Bohr was Oppenheimer's god, so to speak; and this man who was so vain, so arrogant, just hung on every word and glance of Bohr's. He never took his eyes off Bohr for a moment, and it was an amazing and very striking performance. Greatly as I, too, admire Bohr this went far beyond ordinary loyalty, admiration, whatever. Oppenheimer completely subjugated himself in this instance. It was a kind of mirror identification with Bohr, almost as if Oppenheimer was trying to work his way into Bohr's skin.

Do you know Sir John Eccles, an expert on the brain, who believes in the possibility of telepathy and is also a Nobel prize winner?

Wald: I do. I met him lately. Eccles is a very distinguished neurophysiologist, and I'm not surprised about his interest in telepathy. I think ESP research has one fundamental problem: it attempts to scientize what is unscientizable. I've rather changed some of my own attitudes, not towards ESP, but toward the whole fundamental problem of consciousness.

THE MYSTERY OF CONSCIOUSNESS

Wald: You see, I've spent my scientific life working on mechanisms of vision. The work went very well and what was at the beginning a very lonely enterprise now has literally thousands of people working in the field. So we have learned a lot and hope to learn much more. But nothing, nothing whatsoever in the entire package touches or even points in the direction of what it means "*to see.*" With human beings there's plenty of evidence of consciousness, because they use language and so on. But I worked on eyes, what light does to eyes. The creatures I worked on most at first were frogs. Eventually I worked a lot on human eyes. Everything I could find in frogs is very close to what one finds in human eyes. But *I know what I see.* Does a frog *see*? It responds to light, but any number of mechanical devices do that, too: a photocell-activated garage door, for example. Does the frog *know*? Is it *self-aware*? There's nothing I can do as a scientist to answer that kind of question. That means all the talk of "consciousness" tends to take liberties with the realization that it lies outside the boundaries of science; not only present science but, I think, whatever science will accomplish in the future. Strangely enough the problem of consciousness is the day-to-day concern of theoretical physicists. Every top physicist, and that includes the generation that no longer has any survivors, the great generation at the early part of the century, Rutherford, Pauli, Einstein, Bohr, Heisenberg, all of them had come to recognize that one cannot exclude the observer from his observations, that what the physicist, indeed any scientist observes is what he has *prepared* to observe. I think that all good physicists recognize that physics as indeed all science lies within a much wider territory of which we are conscious.

Any assumption I have regarding the presence or absence of consciousness in any non-human animal remains an unsupported assumption. I cannot prove or disprove it by scientific investigation. One presupposition that runs throughout the situation is that consciousness is located in the brain. The reality is that since the presence or absence of consciousness defies all scientific approach we have no way of locating it. But more, just as an electron *does not have* a precise position and motion, I believe that consciousness *has no* location.

Several years ago a thought struck me that at first seemed so aberrant as to embarrass me. This was that mind, rather than being—as I and almost all other biologists have thought—a late product in evolution, dependent on the development of complex central nervous systems, had been there from the start; and that this became that intrinsically very improbable thing, a life-breeding universe because the constant and pervasive presence of mind had guided it in that direction.

That seemed to me a pretty wild idea; but it took only a few weeks for me to realize that I was having it in excellent company. Not only are ideas of this kind prevalent in millenia-old philosophies, i.e. Hinduism; they come up again and again in the works of some of the most prominent physicists of the past generation.

The thought is strange enough: it is that aspects of mind pervade the universe.

I am happy to have come upon that thought on my own. But as I say, I shortly realized that I was in good company. Sir Arthur Eddington in 1928 put it this way: "The stuff of the world is mind-stuff . . . The mind-stuff is not spread in space and time . . ." (i.e., has no location, no extension, either in space or time). I like most of all Wolfgang Pauli's statement in 1952: "It would be most satisfactory of all if *physics* and *psyche* (i.e., matter and mind) could be seen as complementary aspects of the same reality." What is meant is that one has no more basis for considering the existence of matter without its complementary aspect of mind than for asking that elementary particles not also display the properties of waves.

What does it mean to say that matter—all matter—involves aspects of mind? It does *not* mean self-awareness. Self-awareness such as human beings display (and dogs, cats, horses, other mammals) may indeed depend upon the evolutionary development of complex central nervous systems. But what would it mean to ascribe some aspect of mind to a stone?

This is a question that I still puzzle over, and hope to come further with; but let me at least make a beginning with it. I have in the past compared the two elements, carbon and silicon. They share certain important properties, yet differ in a property that is crucial for life. Thus carbon dioxide is a gas in the atmosphere and dissolves readily in all the waters of the Earth—the places from which life on

Earth derives its carbon. Silicon dioxide, however, is the stuff of sand, most rocks, quartz. So I end by saying . . . and that's why silicon is good for making rocks, but to make life requires carbon. That sounds like disparaging silicon. But now let me say something more, that I think begins to approach what one might mean by an aspect of mind in a stone. If silicon were not good for making rocks, there would be no place in the universe in which carbon could make life. For silicon makes the surface layers of plants, the universal abode of life. What I am getting at is the marvelous *fitting together* of all aspects of this life-breeding universe.

Let me give another example: Of the 92 natural elements, 99 percent of all living matter is made of just four: hydrogen, oxygen, nitrogen, and carbon. I think that must be wherever life arises in the universe, because only those four elements possess the unique properties upon which life depends. Now something wonderful: life, to persist indefinitely on any planet in the universe, must come to depend upon the light from its star. So it is that all life on Earth runs on the light of our star, on sunlight. And what makes sunlight? Those same four elements that constitute life on the Earth also makes the sunlight on which that life runs.

A great, past Harvard physiologist, Walter Cannon, wrote a fine, seminal book called *The Wisdom of the Body*. It explained the intricate mechanisms by which all parts of the human body are integrated to further its survival and its multiple activities. One could speak in this sense of the wisdom of the planet, of the wisdom of the universe. When I think of an aspect of mind in a stone I mean neither self-awareness nor intelligence. I mean participation in this kind of wisdom.

How confident can we be in such ideas?

Wald: Let me just say to you that reality is in the head. And now I'm talking present-day physics. If you raise the problem of "the reality of the electron," you'll be told there are experiments that produce certain results which we characterize as "the electron." It has both wave properties and particle properties. To deal with such mutually exclusive sets of properties, Niels Bohr started his principle of complementarity, which accepts the realization that numbers of phenomena, in and out of physics, display and can be completely described only in terms of mutually exclusive sets of properties.

Enter the play of consciousness: you the physicist, starting an experiment on photons (particles of radiation) or on electrons, decide beforehand which of those sets of properties you will encounter. If you do a wave experiment you get a wave answer, if a particle experiment you get a particle answer. To begin to understand the nature of radiation or electrons you need to do both kinds of experiments and live with both kinds of answer.

Let me give you a further example of physics being in the head.

Physicists, as far back as Newton, have been troubled by apparent *action at a distance*, by material objects acting upon one another without any apparent contact. In the last few decades physicists have decided to get rid of the *forces* of physics with which we've lived for so long—gravitation, for example, and electrical attractions and repulsions. Their place is being taken by presumed particle interactions. So it is assumed that every electrically charged elementary particle—every electron or proton—is surrounded at all times by a cloud of photons, and that the electrical attractions and repulsions the protons and electrons exhibit are caused by interactions among those clouds of photons, stronger and stronger the shorter the distances between them. And now something well inside the head: those are not real photons, which would be detectable, but "virtual" photons that cannot be detected because each of them exists for too short a time. They appear and disappear within the time limits of the uncertainty principle.

As for gravitation, that is now assumed to be mediated by the interaction of as yet undiscovered, massless particles called *gravitons*.

In the same vein, physicists no longer think of a vacuum as "empty space." On the contrary, it is filled with ceaseless appearance, interaction, and disappearance of all types of "virtual" particles, all undetectable, since it all comes and goes within the time limits of the uncertainty principle. A vacuum is no longer defined as "empty space," but as the lowest energy state of that portion of space. And as for those interactions . . .

What I'm saying to you, if you have not made contact with it, sounds a little wild. You can read about all of this in Zukav's *The Dancing Wu Li Masters* or Capra's *Tao of Physics*. I'm just trying to say to you, you're dealing in people who have seen ghosts. And present day physics is full of ghosts. Ghosts in the sense of undetectable "entities" whose presence nevertheless gives rise to certain subtle physical effects—not spirits of the dead.

What still keeps physics physics is a context of observations that are highly repeatable.

Would you call Isaac Newton's interest in theology and in prophecy an aberration for a scientist?

Wald: No. Heavens knows, he paid his way. Incidentally, he is my god among scientists. I think the business of science, largely forgotten in this generation, is to try to come to grips with reality, of which science represents by far the smaller portion. I have copies of two of Newton's writings that were published posthumously. One is his study, the *Book of Revelations*, and [the other], *A Chronology of Ancient Times*. Fantastic. It tells you what Jupiter and Moses were doing, year by year. Pretty strange. He was a deeply religious person, as indeed many great physicists have been.

In your 1981 lecture at the Pan-American Biochemistry Congress in Mexico you expressed your belief that mind created our physical universe and ourselves. If you believe this and also admire Isaac Newton's way-out investigations, why don't you, for example, support scientific investigation of telepathy?

Wald: I know very well there are people who take it for granted that ESP has been scientifically demonstrated long since and that there are just a few strange, overconservative holdouts who don't realize this. Also there has been, I think, a more sympathetic audience for this in England than in the States. It rather interested me when I was in England to ask at tea in a Cambridge University laboratory, for example, "Do you believe in ghosts?" Americans, perhaps, because they are more materialistic, more standardized and less imaginative, would invariably shut the question off. But what interested me was that numbers of English scientists would reply, "We really don't know, do we?" One distinguished Cambridge University professor of biology expressed an interest in poltergeists, "One has to be rather careful to sift the good evidence from the poor," he said. "You can't believe all the stories. You have to be rather careful and just accept the ones that hold up."

Well, may I say in my stodgy, materialistic American way, this is all—I suppose I would call it—nonsense. I feel more sympathetic to this kind of nonsense than I used to, in the sense that we are living now in a rather terrifying world. I find it so, and these are aspects of relief at least for the person who wants to get away for a litle while from the terror. The realities of our present world are themselves so blatantly terrifying and irrational and approaching the insane that I feel a little kindlier to these harmless manifestations of the human imagination.

Do dreams interest you?

Wald: I find them extraordinarily interesting. What the hell happens when one dreams? Why does one dream? In one's dreams there's a whole world created and one *sees*—yes? I'm much more conscious of what I see than any other sense. One of the troubles is that unlike my wife and daughter, I almost never remember my dreams. I wish I had come to these interests earlier and had trained myself to remember my dreams.

What are your views on life after death?

Wald: Suppose something does go on after we die, what would be the nature of the thing that goes on? I read the Eastern classics, Hindu, Buddhist, and so on. All of them believe life is a temporary occupation of a body. The body dies, but something eternal and imperishable goes on, eventually to be reborn in another body. It isn't reincarnation but a whole new deal; though paradoxically, an account is kept (karma). I find that Hindu and Buddhist thought of the imperishability, the immortality of what the Hindus call the Self, [soul, or spirit] the Atman, enormously interesting.

I'm in my late 80s [born in 1906]. I'm going to die one of these days. Death has never yet entered my experience. That is I have not in this long life yet seen a person die, or ever been in the same house. I once asked a rather large American audience for how many of them was it true. About two-thirds raised their hands. Death has been taken out of the ordinary American experience. Americans tend to die in hospitals, among strangers, the family waiting outside, being fed tranquilizers until it's over.

Several years ago I first read *The Tibetan Book of the Dead*. To my enormous surprise that feeling of deep resentment that I was going to have to face something of which I was totally ignorant was assuaged. I don't know why I feel relieved but I do.

I once sat with a woman at luncheon who had just come back from a vacation in a hotel that was a spiritualist center. You paid your fee and got all the privileges of a summer hotel plus seances. At the seances, gifts were distributed by the spirits. They came, as well as I could tell, out of the nearest Woolworth's. The vulgarity, the triviality of so many communications said to come from the beyond! For example: "For the spoon that Aunt Mamie lost, look behind the sofa." Are these the concerns of the afterlife?

Do you think that life has a purpose?

Wald: As I said, I began realizing years ago that this universe of ours is a life-breeding universe. It takes no special intelligence or originality to dream up any number of alternative universes which might be fine, stable universes—but lifeless. But we are in an astonishing universe with a special concatenation of properties that makes life possible. There are estimated to be a hundred billion stars in our home galaxy, the Milky Way. At least one percent of them might possess a planet capable of sustaining life. So there are probably at least one billion life-supporting planets in our Milky Way. Telescopes indicate there are a billion other galaxies. So, at least one billion billion planets in our universe might support life.

A few years ago major physicists began to speak of what some called "the anthropic principle." This states that the universe possesses the properties it does in order eventually to produce physicists. [Physicist Anthony Zee expresses it this way: "In the anthropic view, the Ultimate Designer is a tinkerer. He tried one design after another until he found one that accommodates intelligent beings."][2]

Such a life-breeding universe enters a phase of evolution that leads to the independent evolution of consciousness.

Humankind then takes a great place in cosmic evolution, one of transcendent worth and dignity, in which our purpose is to know and create and to try to understand. We are an intrinsic part of the universe. Much of the history of our galaxy is bound up in us. The carbon, nitrogen, and oxygen that forms the bulk of our bodies was made in the deep interiors of former generations of stars that have died. The salts of the ancient seas circulate in our blood. We see this universe not from outside, but from inside: its stuff is our stuff. I once wrote, "A physicist is the atom's way of knowing about atoms." In our knowing, the universe comes to know itself.

Chapter Eight. Arno Penzias

"This world is most consistent with purposeful creation."
—Arno Penzias

Just before World War II started, in 1939, Arno Penzias, then six, escaped with his 5-year-old brother, Gunther, from Hitler's Germany to England, where they waited until their Polish-Jewish parents joined them months later. In 1940 the family landed in New York.

After attending public schools in the Bronx and Technical High School in Brooklyn, Penzias went to City College, from which he graduated with a B.S. in 1954.

He got his Ph.D. at Columbia six years later, studying under Charles Townes. But he had a tough time making it "because I'm a bad test-taker." He calls Henry Foley, "the pillar of my education," Ali Javan, "a brilliant experimentalist who should never have been turned loose on students," and I. I. Rabi, a teacher "who demonstrated the importance of asking questions that illuminate." Penzias decided to become a scientist, "because I wanted to make a living and I didn't think I could make it on my social skills."[1]

He and Robert Wilson won the 1978 Nobel prize in physics for radio astronomy research that provided evidence for the Big Bang theory of the origin of the universe.

Since 1981 Penzias has been vice president for research at AT&T Bell Labs in New Jersey, overseeing a staff of about a thousand. Out of those labs have come talking movies, the transistor, the laser, the solar cell, digital switching, the artificial larynx, and stereo recording.

I contacted Arno Penzias on May 18, 1994, when he happened to be on the way to his publisher with the manuscript of his latest book. It seemed appropriate that as an AT&T executive he should speak with me over his car phone.

When your parents sent you from Germany with your younger brother, was it frightening?

Penzias: Probably the most frightening experience of my entire life. Imagine giving a 6-year-old the job of being a parent to a 5-year-old and not telling them where they're going.

What did your parents tell you?

Penzias: They told me that night basically, "Go take care of your brother." And I had no idea that any adults would ever take care of me. In England, at first, we were put in an orphanage in Bournemouth, I think, and then we were put in a foster home until the blitz started. Then we were sent up to Northumberland. Physically we were extremely well treated. But I was extremely lonely. I think I was a reclusive child. My brother adapted somewhat better. On the boat coming over they were really brilliant. They put him in a different room and woke me up in the middle of the night to go and comfort him.

In Germany had you and your parents experienced overt anti-Semitism?

Penzias: I understood that adults greeted each other with the state-ment, "It's only good the children don't know anything." And I assumed it was a locution people used, I'd heard it so often. I assumed that was how grownups said, "Hello." There were some incidents in Munich. Once I was showing off on a trolley car and I used the word "synagogue" as we passed one. And there were such icy stares. The climate changed. My mother immediately dragged us off the trolley. All of a sudden I learned there was this hidden-rules game—that I could put my entire family in danger by just saying the wrong word, and I had no idea what the words were. It's terrifying. There were times when I said things in public, like I said "national socialist" once because I thought it meant "important." And my parents were very upset and told me not to say that again. There were hidden things. I had the feeling I could somehow put my family in danger, but nobody did anything to us. I remember wanting to join the Hitler Youth and my parents' faces clouded up, and said I couldn't. So I knew we weren't normal in some way, but it wasn't clear what that all meant. Because my father was a Polish citizen we were going to be deported to the Polish border. This was after Kristallnacht. [The Night of Broken Glass—which actually lasted a week. It occurred in November 1938 after a 17-year-old German Jewish refugee, whose father had been deported to Poland, killed the third secretary of the German Embassy in Paris. Ironically, the dead man was anti-Nazi. The Nazis used his death as an excuse to step up their persecution of the Jews. They destroyed and looted hundreds of Jewish stores, set scores of synagogues and homes ablaze, and killed several people trying to escape the flames. Some 20,000 Jews were arrested, and many were deported or sent to concentration camps. The Penzias family was to be deported to Poland but there was a midnight deadline after which the government would not take any more deportees.]

My brother and I were staying with my grandmother and we were taken to prison where we met my parents and cousins. I remember this being a pleasant experience, because we were all in the same large cell and my brother and I and our cousins climbed up and down the bunks. We were then put on a train which took us to the Polish border: the open field where the Poles put the Jews that they were forced to take (by the Germans) and half of them froze to death. Fortunately our train got there after the deadline. So we were returned to Munich and my father was given six months to get out of Germany or be sent to a concentration camp. The people in the apartment we lived in in Munich apparently treated my family reasonably well. A wife of an S.S. man who lived in the building shopped for my mother, because Jews were restricted from shopping. There was however one man alleged to be an anti-Semite and apparently his dog pushed me down what turned out to be a flight of stairs. I saw it as an adult at a subsequent visit and this huge staircase was really just a few steps. I fell to the bottom on a grate and cut my head and got stitches. I still have a scar there.

What happened to your younger brother?

Penzias: He became a chemical engineer, has three grandchildren, lives in New Jersey, and works for the United States government.

Did your parents know of your Nobel prize?

Penzias: My mother, unfortunately, had suffered a stroke three years before. But my father, now 83, went to Stockholm. He loves to talk about me. So just give him a call.

I did and asked: What happened to your boys when you sent them out of Germany?

Karl Penzias: The children went to a home for refugee children in Northumberland. Then I was in a camp near Ramsgate in England. My wife made contact with a doctor in Northumberland and when the children were in the home, she applied to move with the children to the doctor's house where she worked as a domestic. Otherwise she couldn't have come to England. I visited them there and after the war started they allowed us to get together, because I had a ticket for the boat and a visa for all four of us to go to America. We went into the Irish Sea and hit a mine. One sailor was killed and we went back to Liverpool. We stayed for a few days and then made the journey to the United States, arriving on January 5, 1940.

How were you treated in England?

Karl: This was a camp for Jewish people. The British government allowed refugees to go into that camp so we could be safe from Hitler.

It wasn't a prison camp?

Karl: Exactly the opposite. An internment camp for about 3 or 4 thousand people with visas to go to the United States. One of my best friends got me the visas in 1938. He went ahead to the U.S. and asked relatives and friends how he could get us out of Germany through affidavits, which he got.

Had you experienced much anti-Semitism in Germany?

Karl: Very much. I was a leather merchant, selling hides to shoemakers, and I couldn't work any more. At the end of October 1938—

we were Polish citizens through my parents: as Jews we never could become German citizens—we were rounded up in Munich and had to go to the police station where they put us in jail. Next evening we went to the railroad station and were put on a special train to the Polish border. But there was a deadline and we were lucky we arrived too late, after midnight, and they didn't take us. I want to tell you, not everyone was a Nazi. When the policeman who took us on the train heard we could come home he kissed the women. This man had known my parents and all the family for maybe 40 or 50 years. The police were not Nazis. But a lot of boys who went to school with me didn't talk to me because they were afraid, or they tried to break into Jews' apartments and steal anything they could. Anti-Semitism was terrible. There were people, even the best friend I had, would spit on you. I couldn't sell any more leather.

The ordinary Germans, like your neighbors, were anti-Semitic?

Karl: No, I have to say. An SS man, like a major, moved into our building and he never said a word. His wife came down on Kristallnacht, and looked around—nobody could see her—rang our bell and asked if she could bring us food, because we were not allowed to go to the stores any more. And she was very helpful. They did not destroy my apartment, because these two guys said, "We are here. Nobody goes in here." So they protected us.

Those two guys were not Jews?

Karl: No, they were Germans but they knew us from before. Not everybody was a Nazi. Because, if every German was a Nazi not one Jew would have come out alive.

When you lived in Munich did you see much of Hitler?

Karl: Too much.

What was your impression of him?

Karl: That he was a crazy guy. The only thing I must say he was a fantastic speaker and because of this he almost ruled the world.

In Britain when we heard him speaking he sounded hysterical.

Karl: Yes. But he was able to fire the masses so they didn't think any more. Some of the more intelligent people didn't fall for that, but the masses did. In 1932 when he was elected, people didn't have much, and they said, "First we'll try the Nazis and when they fail we'll have the Communists." But he didn't let it happen.

[When they first arrived in the United States, the Penzias family lived in a Bronx apartment. Karl worked as a superintendent for several buildings, collecting garbage from dumbwaiters and stoking furnaces and his wife eventually worked in the garment district.]

What was it like in Sweden when your son got his Nobel prize?

Karl: Fantastic. Everybody there was marvelous. I was with the ambassador to Sweden. I was in a synagogue there and it was funny because when I came in they thought I was the Nobel prize winner. I said, "No. Not me. My son." It was tremendous. Fantastic.

Arno Penzias was pleased when I told him I'd spoken with his father.

Then he agreed to answer more questions.

Have you often been back to Germany?

Penzias: Yes, in the last few years.

How do you feel when you're there?

Penzias: I had regular nightmares about Germans until 1968 when I went back for the first time, and then the Germans were my size. Before that I remember them as being so tall. Trips after that people were a lot smaller, as they tended to bow as I walked through the door. So it got easier.

Did they ever apologize?

Penzias: Not really. There was one guy, I would almost call him a friend of mine in that I know he's recommended me for various jobs claiming I was at that point, "the foremost observational radio astronomer in the United States." This particular guy was a devoted follower of National Socialism in his youth, and was putting me up for awards and jobs and thinking of me as his friend. I even had dinner at his house because he'd heard of this observatory and he is head of an

Institute—I'm on his advisory committee. These folks all thought of me, and I was typically described as the most famous Bavarian astronomer since Fraunhofer. They understand that a lot of German scientists have gone to America to get better jobs. People will send things from Germany to my colleagues here in the United States, saying, "Say hello to my countryman, Arno." Acceptance is the unnerving part.

They know you are Jewish?

Penzias: Of course. They understood I wouldn't eat pork and they will say, "There's no pork in this." In part I do it for identification. I don't think God will strike me dead were I to eat a piece of ham. In fact, I'm almost a vegetarian. But I always refuse to eat things which are not kosher.

So how would you rate anti-Semitism in today's Germany?

Penzias: Low. I've encountered very little. Austria is a totally different case. In Austria you can almost taste the anti-Semitism. Two examples. I went with my wife to "The White Pony," or some such thing in one of the Lehar operettas. There's a jewelry store with a large case of souvenirs. They have huge 5 or 6 centimeter silver coins with the likenesses of famous people. You can buy Winston Churchill, Albert Einstein, and Adolf Hitler. I had another experience when I went to Vienna, Austria. I went to an orthodox Jewish home on a Friday night and normally, while I'm not orthodox, I am somewhat observant. If I go to a synagogue in my hometown I wear a skull cap not just in the synagogue but walking to it, as part of getting in the mood and spirit of the situation. Orthodox people keep their heads covered at all times. Out of respect when I'm going on a sabbath evening to an orthodox home I put on a skull cap before I enter the home. So I did that with the family in Austria. Typically the women stay home on Friday night to make the sabbath meal and the men go to the synagogue alone. So we were about to walk out of the door and this guy got very upset that I was going to wear the skull cap in the street. Here I am, a 50-odd-year-old person with an American passport in my pocket and this is 1990. An American tourist with a skull cap is not safe in Vienna? Ultimately he wanted me to put on one of his ordinary hats. But it wouldn't fit. He finally told me to take off my skull cap, which, for an orthodox person, is going pretty far.

And you did?

Penzias: Yes I did. The point was I put it on as much in deference to his wishes as anything else. And to make him totally upset—to spoil any chance of his enjoying his sabbath, because he would be nervous that I would be somehow attacked—was horrible. People tell me Austrians were somehow able to paint themselves as victims of the Nazis and as people who never did anything wrong. By and large Austrians were enthusiastic citizens of the Third Reich. The only difference was the Germans lost the war and the Austrians didn't. They didn't have to. They got away with it—but on the side of the victors.

Asked how the Nobel prize had affected him, Penzias once said, "[It] can make one feel intimidated. One sort of feels that the people around you are sharper and perhaps a little bit smarter; that they get things faster and don't have all the doubts and questions that you have and that the stuff you understand is, maybe, simpler. I think that's a general feeling among scientists. I don't know about other fields. But I think around here the really good people do have a certain amount of anxiety. Also, when you win the Nobel prize you get into the same company with people like Hans Bethe and Charles Townes, and one can have the feeling that those people are really more deserving. I have sort of finessed that by saying that I don't want to be judged by how much I deserved the prize but rather on what use I have put it to. [For example, Penzias gave his Nobel lecture in the apartment of Soviet dissident and computer scientist Viktor Brailovsky who, in 1980, was imprisoned for "defamation of the Soviet state."] Among other things, you tend to meet a hell of a lot of interesting people. It is, as the sociologist Robert Merton put it, a "haunting presence." If I go to a party with people I don't know, sometimes I think I am having this great time with all these people chuckling at my jokes. And then, after it's all over, people come and say, "It was an honor to meet you"—as if they weren't listening to me at all. So it's a haunting presence you carry with you, but it's not terrible."[2]

What happened to Viktor Brailovsky?

Penzias: Viktor and his family have since moved to Israel where he and Irana both work at the University of Tel Aviv. Their son recently completed his Ph.D. in math.

Are there any scientists the caliber of Einstein, Bohr, and Heisenberg around today?

Penzias: In some sense it was easier in the past. It's easier to make a big splash in the groundbreaking times. Today it's a lot tougher because the ground has been so plowed over, one has to go into very complex and arcane nitches to make progress in physics. There was a unique moment in human history when the foundations of physical knowledge were made. It's like saying, "Are there any explorers as great as Vasco da Gama?" Well, there are no continents to explore. Just to be able to sit calmly while somebody buckles you in a spacesuit just doesn't qualify.

THE DISCOVERY

Scanning the Milky Way with a giant radio telescope in 1964, Arno Penzias and Robert Wilson of Bell Labs were searching for possible sources of static that might impede satellite communications. And when they tuned the receiver to a wavelength of seven centimeters they heard a faint, steady hiss. Wherever they pointed the huge, horn-shaped antenna, the hiss persisted. And so did their puzzlement. They thought the mystery was solved when they discovered foreign bodies in the radio antenna: two roosting pigeons. They removed the birds and cleaned the antenna but the hiss remained.

As it happened, scientists at nearby Princeton University had predicted that if the universe began with an explosion, faint traces of radiation should linger throughout the universe at a wavelength of about ten centimeters.* Told of this, Penzias and Wilson got together with Robert Dicke, P.J. Peebles, and other scientists at Princeton and published a joint paper speculating that the background hiss was the lingering echoes of the universe's violent birth.

What these two astrophysicists had discovered was a cosmic microwave radiation that bathes the earth in a faint glow. Its source appears to be the entire universe. It was calculated that in every cubic

---

* In the late 1940s George Gamow, Ralph Alpher, and Robert Herman had hypothesized the existence of radiation throughout the universe with a temperature of 5 degrees Centrigrade above absolute zero. Penzias and Wilson estimated it to be slightly less than 3 degrees.

centimeter of the universe there are approximately 400 very low-energy photons, or light quanta, left over from the Big Bang. For this they shared the 1978 Nobel prize in physics. (The Russian, Peter Kapitsa, also got it that year for his work on liquid helium.)

Penzias explained how he and Wilson made the discovery: "We were researching the Milky Way when we found more radiation than we could account for, and it turned out upon investigation that this radiation was coming from outside even our own galaxy. There's nothing out there to cause it. That radiation was left over (in the form of radio waves) from the initial explosion from which the entire universe erupted."

In 1929, astronomer Edwin Hubble had confirmed Einstein's theory that the more distant a galaxy from us, the faster it moves. The Penzias–Wilson discovery, some 35 years after Hubble's, seemed to clinch the Big Bang theory that the universe was born through a cosmic hydrogen explosion. That explosion, however, had covered its tracks, leaving what seemed an impenetrable mystery.

For, as astrophysicist Robert Jastrow explained, "We can never tell whether the hand of God was at work in the moment of creation, for a careful study of the stars has proved, as well as anything can be proved in science, that the universe came into being 20 billion years ago in a cataclysmic explosion. In the searing heat of that first moment, all the evidence needed for a scientific study of the cause of the great explosion was melted down and destroyed. The shock of that instant must have destroyed every particle of evidence. An entire world, rich in structure and history, may have existed before our universe appeared; but if it did, science cannot tell what kind of world it was . . . The scientist's pursuit of the past ends in the moment of creation."

Biologist Albert Szant-Gyorgyi, awarded the Nobel prize for discovering vitamin C, said, "If there was a creator, he was not a quantum mechanician, nor was he a macromolecular chemist or physiologist—he was all of these."[3]

And Jastrow, though an agnostic said: "Far from disproving the existence of God, astronomers may be finding more circumstantial evidence that God does exist."

Evidence for the expanding universe, including cosmic radiation and significant amounts of helium around the planets, indicates, Jastrow concludes, that the planets started from one hot, dense, mass.

And "in the face of such evidence the idea that there is a God who created the universe is as scientifically plausible as many other ideas. Yet scientists—often out of their own ignorance—tend to ignore God as an explanation for the beginning of the universe. Science cannot bear the thought that there is an important natural phenomenon which it cannot hope to explain with unlimited time and money. There is a kind of religion in science . . . the religion of a person who believes there is order and harmony in the universe. It would violate such a person's belief to conclude that the universe could have begun in a way counter to physics."[4]

Penzias agrees, saying: "The best data we have are exactly what I would have predicted had I nothing to go on but the five books of Moses, the Psalms, the Bible as a whole," in that the universe appears to have order and purpose. "Though I would have added the explanation that when one talks about purpose and order, this is exactly the world one wants to go to. Taking the parameters on the basis of open universe versus closed, pulsating matter versus antimatter, high entropy versus low entropy; of all the exclusions, if one has to pick one that goes with order and purpose, this is the one you would pick."

I asked Penzias: By "the world one wants to go to," do you mean the world described in the Bible?

Penzias: It reflects the same world view, rather than exactly the same world. It's consistent with the same world view, though not exactly the same, in the sense there is not that kind of description. The Bible talks of purposeful creation. What we have, however, is an amazing amount of order; and when we see order, in our experience it normally reflects purpose. That's as far as I can go.

And this order is reflected in the Bible?

Penzias: Well, if we read the Bible as a whole we would expect order in the world. Purpose would imply order, and what we actually find is order.

So we can assume there might be purpose?

Penzias: Exactly.

What do you mean by "this is the world one would pick"?

Penzias: What that means is there are so many possible universes that we could, for example, expect an oscillating universe. It would be perfectly reasonable because you could build a world in which the universe oscillates between sequential stages. We find this is not the case. Nothing else would change in the world other than the universe would be closed; which means that it collapses and explodes, and does that many times. But we find this is not true.

When you say, "This is the one you would pick," is that because it's a world with order?

Penzias: This world is most consistent with purposeful creation.

Do many astrophysicists share Fred Hoyle's view that a larger world exists beyond our Big Bang-created universe, in which some parts expand, while others contract? And where time goes on without beginning or end, in a kind of cosmic perpetual motion?

Penzias: All I can say is we are free to assume that beyond the limits of our observation are things we haven't seen yet. We are free to speculate about anything which we have not directly encountered. In other words, I am free to speculate that tomorrow morning it will be safe to stand in front of moving buses. Okay? The thing is, there's no way to prove that *today*. The point is it's merely uninteresting.

How do you think Hoyle arrived at that theory?

Penzias: Well, people are uncomfortable with the purposefully created world. To come up with things that contradict purpose, they tend to speculate about things they haven't seen yet, like missing mass, which would allow the world to collapse back on itself. Otherwise it's going to expand for ever. In other words, it's people who want to be in cyclical universes who say mass has to be missing so it will collapse on itself again.

And that missing mass hasn't yet been discovered.

Penzias: You say, "yet," and it implies it has to be. It isn't that it is missing, although we say it's missing, as if the astronomers hadn't found it. Whereas, the truth is, it's *absent*.

Dr. Penzias was even more emphatic in response to another interviewer: "The search for all this so-called missing mass is an act of desperation by physicists who are unwilling to accept the principle that in the universe, what you see is what you get. All the evidence suggests that the universe is open. Ultimately the universe and everything in it, created in a blinding flash, is doomed to disappear. There really is only one shot. The universe . . . is a paradigm of our lives— a very definite beginning with an indefinite but inevitable end."[5]

Is it plausible that we live in a world of many universes?

Penzias: No.

Since your discovery of cosmic radiation has anything been discovered to refute this idea of an ordered universe?

Penzias: The observations of the past quarter of a century have strengthened the picture. The open Big Bang universe, which has been around for half a century, is getting more and more adherents all the time. Sure.

Do any major scientists disagree with the Big Bang theory?

Penzias: The real problem is that most scientists don't agree with the standard open Big-Bang theory: a universe created out of nothing with an infinite extent the moment it is created, and has been getting bigger ever since. Just because something is infinite doesn't mean it can't get bigger (surface of a balloon). The trouble with that is, there can be only one universe. On the other hand, if you had a closed universe you can have multiple bangs; with a closed universe it comes back (the crunch). The nice thing about that is, then there is no single event to explain. Something has always been here and occasionally bubbles. Since science can never explain anything, but can only describe, you can describe this phenomenon but never need any explanation—because it's always been here. If somebody says, "How come all your secretaries are women?" you say, "It's always been that way." It doesn't need an explanation. In the same way, if we say, "Why is there a universe?" and say, "Well, it's always been here," one doesn't have to explain it. On the other hand, if we have a universe created out of nothing, and of infinite extent and the only one that's going to be here for ever, that presumably is a little uglier from a physics point of view.

Moreover, the universe, which is infinite, has an excess of matter over antimatter, which would also require more than just an accident. One can understand the clever thing with Higgs bosons, that you can have more matter than antimatter accidentally *because* of a decoupling and a decay. But, since that's an accident, it would be remarkable that regions which are all out of causal contact with one another had exactly that same accident everywhere. So here again is an ugliness in the open Big-Bang theory, because somehow the universe was created with an excess of matter over antimatter, (see chapter seven, George Wald conversation) for which scientists have no explanation. The nice thing about inflation (see Weisskopf conversation) is that in fact there was just one accident. And all the pieces of the universe we encounter now were in causal contact early on before inflation got them out of causal contact, and then coming back into it. So that inflation takes care of the excess of matter over antimatter. But inflation requires that the universe be in fact exactly balanced—with as much kinetic as potential energy—because then you have a zero net energy and creation out of nothing, which doesn't violate conservation of energy. Even though, as astronomers we see a universe which is clearly open by a factor of 100 to 1 at the present time. Remember, though, this is not an inventory.

People talk about "the missing mass," as I mentioned before, as if astronomers haven't counted it yet. Whereas in fact all the measurements of the mass of the universe are questions asking the galaxies what pull they feel one on the other. These are dynamic measurements, not inventories; or the clusters of galaxies which are measured by gravitational lensing. You can measure their mass in other ways which don't even require dynamics. So astronomers believe in an open universe. Physicists on the other hand don't, because it violates their notion of what is an acceptable model of the world—so they throw out the observations.

Do you agree with Robert Jastrow that science seems to confirm the Bible? (Just in the creation of the universe)

Penzias: I wouldn't say it confirms the Bible as such. In a world that has purpose one would expect order, one would expect asymmetries, something other than chaos. In fact, we find a world which is quite highly ordered. So it seems to me to be a necessary but not a sufficient argument for the kind of things that the Bible, or Western religions, talk about: that is, a world having purpose to it. I can't claim to have discovered purpose. All I can say is I discovered order. A perfectly symmetric world is in some sense a disordered world. The asymmetries, the lack of perfect symmetry, that there is more matter than antimatter; the fact that the world is here now and wasn't there before. A perfectly symmetric world is one which is just dead empty, for example. (See chapter seven, conversation with George Wald who discussed this subject with Einstein.)

I thought Einstein believed in an ordered world, but found the expanding universe theory irritating.

Penzias: Okay. He believed in a symmetric world, order in a different sense. Order in a purposeless sense. That is, if you have a world which is always sitting there, it has a much more aesthetic appeal. That is in some sense an appeal to scientific symmetry; a world which is always in existence is more symmetric, is rather nicer. It doesn't have any things that one has to explain. On the other hand, a world which has some very definite beginning at some peculiar time needs an explanation. Very often if we say something however unjust or stupid, we can always argue that we've always done it that way. It's an argument. So, in the same sense, if someone asks a scientist, "Why is the world here?" he can hide the fact that he has no explanation by saying, "It's always been here." So, the scientist who believes in order believes in a very specific kind of order, an order that doesn't require an explanation.

And the expanding universe?

Penzias: The universe expanding from a specific origin is not symmetric in the sense that it isn't balanced by something else. It all of a sudden happened. So that is order of a different kind. It is a progressive order, an order in time. Something which started at a certain time and is going somewhere. The words symmetric and asymmetric are a little ambiguous in this context. And what I think of as order in the world is a temporal order, not totally temporal, but an order of the kind one would associate with purpose. There is another kind of order which one might associate with lack of purpose, a meaninglessness, the kind of thing Steve Weinberg talks about. (Winner of the 1979 Nobel prize in physics, Weinberg concluded there's no purpose in the universe.)

But didn't you once say that the description of creation in the Bible may parallel what in fact happened?

Penzias: When I spoke of the Bible, I was speaking rhetorically and not in detail. I perfectly understand that one doesn't have light, and then the sun next, and so forth. I'm not asking for details. When we talk about Western religion, to make it a little more general, we talk about a world with a beginning and with purpose. Which is precisely now what we find. A world which seems to be what we call the standard, open, Big Bang universe; a single explosion destined to go on forever.

Others who have been trying to solve mysteries by using scientific methods are the parapsychologists. Do you agree that they have tried to test in the laboratory some of the biblical miracles? And that this is a reasonable pursuit?

Penzias: I disagree. The world I see as a scientist is one in which the laws of physics, so far as I can tell—and, of course, there will always be things we only know in an ad hoc sort of way—provide an adequate description of the details of the universe. That is not to say that one is absolutely sure we understand all the laws of physics at any one time. But the kinds of things that make me believe in purpose, or in the Bible, as it were, have to do with the miracle of existence, and not whether somebody can figure out a way of having ten percent better odds at blackjack. Which is what some of these parapsychological things come down to. I wouldn't want ever to connect the two at all.

So what is your verdict on parapsychology?

Penzias: Because they've not been able to produce any experiments which a group of respectable scientists are willing to put into science makes me a little suspicious. Maybe the scientists are so totally blind that they won't look, but I suspect the methods can't be all that good. So my attitude is like the old Scottish verdict in law—not proven. And the extrasensory occurrences and positive parapsychological experiments are rarely ever repeatable.

Nor are astronomical occurrences.

Penzias: But two people can make the same measurement. If I see a star or comet explode and no one else has seen it, that kind of astronomy doesn't go very far. It's the things other people can see and measure that then are given some credence. When you have a situation which depends on the person having some unknown property, you end up with a lot of wishful thinking. It's fairly easy even in objective science to be fooled, but the question of dishonesty comes up in this very quickly, and the dishonesty may not even be conscious. I remember an example of spectroscopy where people nearly knew the answer. People got incredibly persuasive results and that had to do with the fact that they had a certain amount of feedback. It's very hard to eliminate such things. Uri Geller, I suppose, fooled many people for a long time, as an example.

I said to Louisa Rhine, the parapsychologist, "How persuaded are you that ESP exists? Are you as convinced that it exists as firmly as you know that if I chopped my arm off it wouldn't grow again?" And she said, "Yes."

Penzias: Yeah, okay. I'm sure. But remember, she had this tremendous desire to prove the existence of ESP. We've occasionally had people (who claimed paranormal powers) come to Bell Labs and describe various things. And the statistics are way down. Statistics are very, very funny things. I don't think ESP is impossible, but there are so many chances of fraud.

So you're skeptical.

Penzias: Basically I'm very much aware of the limitations of science and of scientists. Part of this comes from human attitudes. The old victory of science over religion, for example, led to certain scientific attitudes. Attitudes of all kinds color our perceptions, so I tend ultimately toward skepticism. In this same spirit, when there is a strong personal involvement in a certain proof, I'd tend to stay away from it.

Einstein at one time seemed interested in telepathy.

Penzias: While Einstein was a great man in many things he was far from perfect. He was human and made mistakes, and had his ups and downs. If some well-known mathematician I respected was able to repeat the experiments, and this was published in *Science*, or *Scientific American*, I would go along with it. If somebody had an honest to goodness double-blind experiment one would have to look at it.

Do you think it's reasonable to study telepathy?

Penzias: It's a perfectly plausible and respectable activity. As long as people are honest about what they do I've no reason to criticize them. But as for trying to duplicate the biblical miracles in the laboratory . . . The Bible very strongly talks against magicians, astrologers, and mediums.

But surely as a scientist you don't believe everything in the Bible?

Penzias: I might say I believe in God but I don't believe in astrology. I think the two are very different. I don't believe in an old man with a beard, but I believe in a well-ordered, and purposeful universe that doesn't need capricious violations of the laws of physics. To put caprice into it seems to be contrary.

How about the so-called miracles in the Bible? Dreams that came true, people reading others' minds, water into wine?

Penzias: There are easier explanations. Any magician worth his salt can do that. We understand that man has imagination and the ability to perceive things in many ways. We have all kinds of non-verbal, non-visual clues which we unconsciously operate on. The kind of world I believe in, which is an ordered and purposeful world, doesn't need capricious violations of the laws of physics.

Do you agree with Francis Crick that "very few astronomers are prepared to give any credence to UFO sightings, if only because the percentage of obviously false reports is so high"?[6]

Penzias: Right. The point is negative proofs are impossible. It is very hard for you to prove you weren't thinking about ice cream the day before yesterday.

Crick also believes, "the myths of yesterday, which our forbears regarded as the living truth, have collapsed . . . Yet most of the general public seems blissfully unaware of all this, as can be seen by the enthusiastic welcome to the Pope wherever he travels."[7]

Penzias: That's complete scientific arrogance. I would say exactly the opposite is true; that the myths of the past have become the observational realities of today. And yesterday's observational realities have turned out to be incorrect. Let me give you an example: Maimonides. He talked about a created universe. Aristotle talked about the eternity of matter. So that religious belief talked about a created universe and it flew in the face of alleged observation. Faith said one thing and experiment said something else. Experiment said matter is eternal. Aristotle and others said so: you look at a rock and you come back next week and it is still there. If matter is eternal, the created universe has to be wrong. Maimonides said, as he does in *The Guide to the Perplexed*, "Take notice of the words of any man, for the foundation of our faith is that God created the universe out of nothing," and so forth. Okay? Now what do we have today? Today we have an observation that the universe was created out of nothing— because that's the easiest explanation for the open Big-Bang universe. Now what do scientists like Francis Crick say? "Oh no, the universe has to be eternal," in the face of observation. In other words, they are not being objective. Their religion is: There has to be a purposeless world. And if the world is purposeless, these observations have to deny observational evidence. It's the scientists who are clinging to "religion," in the face of observation. In other words, the myth of purposelessness, which is contradicted by observation, is clung to by the very scientists who are sneering at the Pope.

When you speak of religion you don't believe in a God with a beard, do you, but of some possible intelligence?

Penzias: I don't know.

You're open-minded?

Penzias: Yes. The point is, one of the things that Judaism teaches us is not to anthropomorphize God. Because Michelangelo painted him with a beard doesn't give us the right to give him one. We have to use the word "him" there in a very loose sense. We are warned in Judaism against idolatry, and we are warned against giving our deity physical attributes. Even though there are statements like "the finger of God," there are long Jewish tractates saying it is not to be thought of in that way.

Are you more of a believer in a religious sense than Einstein was?

Penzias: Oh, sure.

Are you a practicing Jew?

Penzias: Yes.

[Penzias once pointed out that although his "religious faith may color his thinking about cosmology, "If someone disproved the Big-Bang theory I wouldn't start cheating or lying. My faith is not dependent on physics."][7]

I was speaking with someone whose father was an orthodox Jew, who went into oil exploration, oil drilling, and studied geology. Because what he learned contradicted what he'd been taught about his religion, he dropped his religion. Why hasn't that happened to you?

Penzias: Oh. It's like saying, "I have a friend who drove into a brick wall, why go in your car?"

Is that a good analogy?

Penzias: The politest thing I can say about that is, it's pretty naive. The question is: why did he expect to find God in an oil well? That's the question you should ask.

No, he wasn't expecting to find God there, but he found great discrepancies between reality and what religion taught.

Penzias: If he thinks religion is what he learned in Sunday school . . . Should you give up on life if you get to be 8 years old and you find there's no Santa Claus? Should that mean your parents are liars?

You mean there's a much more sophisticated way of looking at religion?

Penzias: I would hope so. I mean, Maimonides was a physician and talked about psychosomatic disease in the terms of Jacob wrestling with the angel. Do you think he didn't really understand the stuff, or he wasn't orthodox? These people weren't fools.

In Francis Crick's recent book on the brain he suggests everything about the mind can be explained physically.

Penzias: That's another form of religion, isn't it? What he's doing is making an unprovable assumption based on his observations. Ought we to credit a statement that says: I know everything about the brain and I can explain it all? Are you serious?

Does he say that?

Penzias: How else can he say everything about the brain can be explained in terms of physics?

I think he would put it like this: Of all the research and reading I've done about the brain, there's nothing that can't be explained physically.

Penzias: That's not quite as meaningless a statement, but it's also meaningless.

Because so much is undiscovered?

Penzias: Yes. Also, wait a minute, that doesn't explain anything about science, right? Science, after all, chooses between explanations. There is no such thing as an unexplainable event. You can explain everything in terms of the physical and always could. It means that at some point you make them too complicated, right? At some point the phlogiston theory was overthrown and somebody came with oxygen. But that doesn't mean the phlogiston theory couldn't explain all phenomena. Of course it could. It just got a little too complicated, that's all. That's what we're really talking about. And for Crick to believe that theories are ever disproven is somehow naive. I would prefer a statement which is that theories are accepted or abandoned; they're not proven or disproven. What he really has said is: I can craft a set of theories which are acceptable to me, which describe the observations I have read to my satisfaction. What he's doing is giving us his reaction to the physical world, and he's entitled to that opinion. He's telling us what he feels, and that's all he's done. Now scientific intuition, especially from someone as gifted as Crick, is worth something. After all, all we ever have to go on is scientific intuition. And the great miracle of science is that intuition actually helps us and guides us in a sensible way.

Intuition and math, eh?

Penzias: Well, nooooo! Mathematics is just a tool to guide our intuition. Math isn't separate, it's just one of the tools. It turns out as Kepler, the biggest true believer, said. He thought God was going to be a mathematician and it turned out to be a very fruitful supposition.

You remember someone said at the end of the 19th century that there was nothing left to discover.

Penzias: In classical physics there wasn't. So there was a new field that went beyond it. And remember that particle physics and superstrings, and all this stuff we're talking about is a rather high-energy niche in quantum physics. Nobody doubts general relativity, for example, and some of the stuff is corrections to quantum mechanics, after all. The problem is that the area in which the people are working now doesn't lend itself to tests in the same way as the work of the past, because they're so much further away from human experience. So it's just a tougher arena.

Like the fraction of a second after the Big Bang.

Penzias: Yes. We're talking about things that have lifetimes of one zillionth of a pecosecond—and fantastic energy. So we're learning about little resonances which is interesting but hardly the same as discovering the first half-dozen elements.

Any special way you get ideas?

Penzias: Not sure. Not clear. Dreams don't do it. What would I say? Things sort of percolate. I often think with a piece of paper in front of me, a yellow pad with a blue pen and blue lines on the pad. Concentrating sometimes works. Other times it comes when I'm reading something, or from a conversation.

Any interest in hypnosis?

Penzias: No.

Why not?

Penzias: Not enough time I suppose.

It's a fascinating subject. Through it you might dredge up forgotten memories vital for your work, for example. Do you have a lifetime goal?

Penzias: To make the slow buck respectable. That is to say, I'm trying to make the doing of research, the adding of intellectual value by the commercial enterprises, profitable. I'd like very much to add value to our society. Instead of throwing people out of rent-controlled apartments and into condos, I'd like to see people creating new value which would give others employment and a better life through research. This has been my career for the last decade or more.

What book are you working on?

Penzias: I've just finished one and am about to take it to my editor and drop it in his lap. The provisional title is *Harmony*. The subtitle is *Business, Technology, and Life After Paperwork.*

Are places like Russia and the Ukraine asking for your help?

Penzias: Oh sure. In fact the city my grandfather came from which was originally in Austria, then in Poland, then in the Soviet Union, and now in the Ukraine, once called Lemberg and now L'vov, got the first digital switch from AT&T, the first one delivered in the former Soviet Union. So we are putting in switching equipment in various parts of the Ukraine and doing stuff with Russia as well.

What do you think of accusations that Oppenheimer, Bohr, Szilard, and Fermi gave Russia information about the atomic bomb? (According to a Russian official.)

Penzias: I put it in the same category as people who claim to have had sex with an important political figure just after they've signed a book contract. It's not surprising. I tend to be very suspicious of things of this sort. These files are always full of disinformation, anyway. We've strayed very far from a system of justice, where people are in courts of law, and to assassinate people's reputations with the slimmest of innuendoes is unworthy of the profession. You can read in the proceedings of the Security Council of the U.S., Adlai Stevenson testifying that the U.S. had nothing to do with the Bay of Pigs invasion. Nobody thinks that John F. Kennedy, one of our heroes, lied. At that point they found it was in their interest to lie: historical evidence. Whether the charge against Oppenheimer and the others is plausible or implausible, it's unproven and I don't think we should be speculating on things like that. It's not a useful use of history.

Did Oppenheimer get a fair deal?

Penzias: Probably not. If I think of the hysteria at the time, the red-baiting and all the other things that went on, I doubt it. I can't imagine that someone as intellectual as he was—not hail-fellow-well-met, he had enemies—I can't imagine that such a person gets a totally fair deal. Now, whether he was innocent or guilty, I have no way of knowing. Whether he should have kept his security clearance or not, I'm not sure. But I don't think it was totally fair.

Was Star Wars viable?

Penzias: I was skeptical.

What do your three children do?

Penzias: My son, David, is a marketing vice-president for a computer company. My middle daughter, Mindy, is studying for a Ph.D. in clinical psychology in California. Both are married. And my youngest daughter, Sifshifra, is studying to be a rabbi.

Any career disappointments?

Penzias: The great disappointment is that AT&T was broken up and Bell Labs isn't what it was. And America doesn't have the position of unquestioned leadership in the world. Things of that sort. Of personal disappointments, I wouldn't say so.

What do you read?

Penzias: *Wall Street Journal, New York Times.* I skim magazines on airplanes rather than read them regularly. I don't read scientific journals. I read preprints, reprints, technical reports, newsletters, books. I'm reading *Lend Me Your Ears*, William Safire's great speeches in history. And a science fiction story. That's my nighttime reading.

Are you a science fiction buff?

Penzias: No, I read it because my son-in-law who is, gave it to me. It's a sequel to a book, *Ender's Game*, which I thought quite good. The story of actual war played as a video game, and this is a convoluted sequel which I'm not as pleased with and probably won't finish.

Do you watch anything on TV?

Penzias: No, I don't have a television set.

How about hobbies?

Penzias: Writing, hiking, swimming, skiing. Physical things. I used to craft kinetic sculpture.

Any interest in music?

Penzias: Not serious interest. I can't read music, for example. For easy listening kind of stuff, baroque mostly. Vivaldi isn't involving enough. Bach, perhaps. But I never listen to music when I'm writing. It's too distracting. I get too involved in the music. I don't treat it as elevator music. I actually hear it, and then I pay attention to it rather than working.

If you were 18 today what career would you choose?

Penzias: Probably biology. It's probably the most open and interesting field.

How about neuroscience?

Penzias: I'm less interested. I don't think it's going to go as far. They've done some interesting easy stuff, but I don't think the ability to take the brain apart will come any time soon. Whereas I think if I were 18 I would say the coming decade, where I'd be getting my Ph.D., would probably be the years of progress. In the sense at that point being able to map and manipulate the macromolecules which govern life. That's very exciting.

Chapter Nine. *Robert Jastrow*

"In countless solar systems . . . science must have . . . created a race of immortals that may be heading for our sun."
—Robert Jastrow

If it hadn't been for rats, Robert Jastrow would have been a doctor. At 16, in 1941, he was a premed student at Columbia University. His psychology professor, Fred Keller, suggested calculus as an efficient way of tracking the exponential learning curves of rats. So Jastrow took a course in calculus and was hooked. "I loved math and my life changed from there on," he said. He has no regrets about abandoning a medical career, "because I enjoy theoretical physics enormously."

Jastrow was founder and director of the Goddard Institute for Space Studies at Columbia University (He called it the "long-haired part" of NASA), professor of geology and astronomy at Columbia, and of earth science at Dartmouth. He is presently director of Mount Wilson Observatory in Pasadena, where the search is on for extraterrestial intelligence among the 1,000 or so nearest stars as old or older than the sun.

"Very few scientists write as fearlessly and honestly as Dr. Jastrow," said Bernard Lovell, director of the Jodrell Bank Observatory. His provocative books include *God and the Astronomers* (W.W. Norton, 1978), in which he wrote, "Far from disproving the existence of God, (astronomers) may be finding more circumstantial evidence that God does exist" and *How to Make Nuclear Weapons Obsolete* (Little, Brown & Co, 1985), a vigorous endorsement of SDI or "Star Wars" research.

In his fourth and favorite book, *The Enchanted Loom: Mind in the Universe* (Simon and Schuster, 1981), he explored the creation of the cosmos, the origin of life, the development of man's brain, and predicted life in the distant future that makes most science fiction seem tame. Jastrow suggests that "in countless solar systems older than ours, science must have . . . created a race of immortals that may be heading for our sun." As director of California's Mount Wilson Observatory, he can look for those immortals through a 60-inch telescope.

I spoke with Robert Jastrow on February 9, 1994, a few days after an earthquake. "The whole observatory at Mount Wilson was undamaged and my apartment home, but I was terrified," he said. "It was the most frightening experience I've ever been in. Plus there were 20 stories of steel and concrete bouncing up and down over my head. It lasted 40 seconds—like an eternity, really violent, thumping and banging and moving up and down—the whole building. It was awful. It was only a 6.8 but the vibrations were very threatening."

After the Mars probe you said: "It looks like there is life on Mars." Does Gil Levin of the Viking team share that view?

Jastrow: I share *his* view. He did the experiments. Microbes, at least. Because what Levin looked for was microbial life. Several people have come around to agreeing but NASA never got into any arguments about it and it never got much public notice. Incidentally, one doesn't do science by taking votes and arriving at a consensus. The evidence is clear, and I'll stand by it. I don't care who agrees with me. Ponnanperuma, a biochemist at the University of Maryland, is very interested in pursuing the matter.

Do you often study Mars from Mt. Wilson?

Jastrow: Not personally. We do have a project underway to take advantage of the clear air on Mt. Wilson and the sharp images. Nothing more can be done on the probe until we get back to the surface of Mars. I'm most interested in observations that shed light on the earliest moments of the so-called Big Bang, which might lead to a clue to this question on the borderline of philosophy and science, which is: What caused the Big Bang? In the broader context, which I personally am interested in, of not only physics but life in the cosmos, I look forward to the possibility of confirming that there's some kind of life, no matter how primitive, on Mars, or some indication of life elsewhere. Because, at the moment, we have no idea whether this miracle of evolution of life out of nonliving matter, while scientifically explainable, is such a small a priori probability that it has only happened once, here on earth. Although I believe there are microbes on Mars, I have to get someone else to agree with me and that will require going back to Mars.

Does that biblical account in Ezekiel which you suggest may describe the landing and takeoff of a spacecraft, strengthen your hunch that there's life in space more intelligent than us?

Jastrow: I consider that a throwaway line. It's a nice point. If you've looked at that passage. ["Behold, a whirlwind came out of the north, a great cloud, and a fire infolding itself, and a brightness was about it, and out of the midst thereof as the color of amber, out of the midst of fire. Also out of the midst thereof came the likeness of four living creatures . . . they had the likeness of a man . . . And the living creatures ran and returned as the appearance of a flash of lightning . . . And when the living creatures were lifted up from the earth, the wheels were lifted up . . . And when they went I heard the noise of their wings, like the noise of great waters." Ezekiel 1:5–24.][1] It really is very convincing, if you want to be convinced.

You called us earthlings worms compared to the creatures who might have evolved in much older planets.

Jastrow [chuckled]: What I said is that on the average if the universe is 15 billion years old, as I believe—despite a recent estimate of 10 billion years—then stars on the average are $7\frac{1}{2}$ billion years old, which is 3 billion years older than we are. And if life proceeds more or less steadily toward complexity, as it has on this planet, they are also more than 3 billion years more evolved. And if you ask what 3 billion years means in evolution, I point out that a billion years ago on the earth the highest form of life were worms.

What do you think of UFO reports?

Jastrow: They're flawed in the sense that any creature that could physically transport itself across interstellar distances is bound to be much more advanced than we are and such an advanced creature wouldn't come down in the forests of New Hampshire and pick someone up and give them physical exams. [I don't follow this argument. We're more evolved than microbes, but don't we pick them up and examine them?] I don't think they're fraudulent, but they may be self-deluded. If anyone actually does project his physical presence across those distances it must be from one of the advanced societies, and the likelihood of their being as close to us as the creatures in "Star Trek" is small when you consider we've been on this planet for about 100,000 years as the defined species and on the average beings around us are billions of years younger or older. So the chance of being just in that particular notch of homo sapiens is almost vanishingly small.

Do you believe humans have extrasensory power?

Jastrow: My mind is open to all possibilities of this kind. Except I think it's not likely any human being has those powers yet. I think they exist around us in the cosmos, but not on earth yet. But that's an unsupported opinion.

You mean there are intelligent creatures elsewhere who have these powers?

Jastrow: Yes. Knowing we were very young compared to the age of life in the universe, I assume the way we communicate is very primitive and that there are other means.

I tried to get your fellow astrophysicist, Arno Penzias, to support your view on the plausibility or possibility of God. The possibility he goes along with, but he said he meant it rhetorically.

Jastrow: The thing is there was a beginning to the universe and Penzias has proven that that event may have been caused by physical forces that we would recognize. Or it may have been something we would call supernatural. But the point is we can never tell. The circumstances close the door on this question, at least to science. And that's interesting.

Would you call the Bible, at least metaphorically, a blueprint for creation?

Jastrow: There's only one element of the Bible that coincides, and that's the evidence for the abrupt beginning. The first verse, "In the beginning God created the heaven and the earth." And from there you can go off each in his own direction.

Both science and the Bible now agree that the universe started in a flash of light and heat, an immense explosion. Right?

Jastrow: Well, a flash of light and heat.

In *God and the Astronomers* you wrote that because of the obscuring fog of radiation, it's not possible to see anything that happened in the first million years of creation.

Jastrow: I remember writing that. Before the first million years you just have these negatively charged electrons and positively charged nuclei, and they make an electrical plasma that is opaque to light. So the sense of that is that light cannot travel an appreciable distance so you therefore can't see that far. I think I have to stick with that.

Does quantum mechanics makes it possible to determine the state of the universe a fraction of a second after the Big Bang?

Jastrow: That's sheer speculation. That's something like asking how many angels can dance on the head of a pin. Because you have little or no actual observational evidence about the universe that early. The earliest direct evidence is about three minutes after the Big Bang.

Water appearing a second after the Big Bang and the temperature dropping a billion degrees is again speculative?

Jastrow: The condensation is much more extreme than the numbers you mention. And it has to be taken seriously, because it does explain a few things in cosmology. But it seems to me to be a picture of the universe so contraintuitive I would reserve judgment.

How about atoms appearing after one million years?

Jastrow: Yes. In the first 30 minutes or so, neutrons and protons came together from helium nuclei. Then nothing much happened until about a million years had gone by, then electrons were captured by these nuclei to form atoms. And then nothing happened for about a billion years, when stars and galaxies began to form.

Penzias says the universe appears to have order, and from order you could assume there might be purpose.

Jastrow: Oh, really? Then he's a Deist, I would say. I can't say that. I can't say it doesn't. I think privately it does not: that the order and purpose somehow are expressed in the laws of evolution.

Does astronomy confirm the 20 billion years for the age of the universe and 5 billion years for our earth?

Jastrow: There's a relatively acrimonious dispute between those who say its 20 and those who say its 10. I usually say 15. But I think the 20 billion fellows are winning.

Were you misquoted when you said, "Everyone in the scientific community is convinced life is common in the cosmos."?

Jastrow: I shouldn't have said everyone. But there's a tendency among scientists to believe life is common on the basis of what's called, "the principle of mediocrity," which says the earth is made of materials common throughout the universe and as an ordinary planet around a very common type of star, why should only this planet be blessed with this event?

With billions of possibilities.

Jastrow: That's a great underestimate. There are 100 billion stars in our galaxy of which one billion or so are stars similar to the sun, and there are 100 billion similar galaxies within the limits of the observable universe. There are a lot of stars and planets out there like ours.

Why was your telescope mothballed?

Jastrow: In 1985 it was closed by Carnegie to put its resources into other projects. We opened it a few months ago. It's fully operational now. We're about to announce to the community that time is available on the 100-inch telescope.

Is there a precise figure of how much clearer objects would appear seen through a telescope above the earth's atmosphere?

Jastrow: Yes. And we've achieved that on the ground with Star-Wars technology. But the answer is that Mt. Wilson has the best seeing on the North American continent, according to Horace Babcock, because the inversion layer over Los Angeles suppresses turbulence at our altitude and these sharp images correspond with the fact that a large fraction of the time the so-called "seeing" is better than one arc second. If you had two objects a mile apart on the moon, an arc second is roughly how far apart they would look from the earth. So it's a very small angle, and most sights in North America don't do as well. If you put on this telescope—if you attach to it the instrument developed by SDI for keeping a laser beam focused as it travels through the atmosphere, you get to within 20 percent of the sharpness of image which you would have if the telescope were in space. The 60-inch telescope on which we've mounted this instrument,—about one tenth of an arc second. We are designing and about to construct a system of so-called adaptive optics that takes the twinkle out of starlight, and to put it on the 100-inch telescope. Hope to have that done within the year. That will give us images with the sharpness of about one-twentieth of an arc second, which is about the same as the Hubble telescope in space.

How far can the Mt. Wilson telescope peer into space?

Jastrow: When Hubble was using it, it could go to about one quarter of the way out, speaking loosely, to the edge. Now, because of the lights of Los Angeles we can't look so far out. We can just barely see to the other nearby galaxies and we do most of our work on stars in our own galaxy which are unaffected by this problem.

Do other telescopes probe much farther?

Jastrow: Yes. Hawaii has darker skies, and down in Chile where the skies are considerably darker the Carnegie Observatories and Europeans both have telescopes.

In the next ten years might you find anything as remarkable as Hubble's evidence for the Big Bang and the expanding universe?

Jastrow: I don't think that will happen. But we may find direct evidence of planets in other solar systems, some close to our own size, and that would be important. We think they're there, but we don't know yet. There's a program with the 60-inch telescope designed to study stars similar to the sun and look for variations in their energy output, which are known to be correlated with variations in the so-called sunspot activity. We find, while the sun itself is in a fairly quiet state at the moment, stars like the sun undergo large excursions of energy output, of brightness. If they occurred in the sun, it would account for all the recent climate changes we have experienced. And that would be of great importance. Then, Charles Townes, codiscoverer of the laser, has a large facility set up on Mt. Wilson. It's an interferometer, which works at infrared wavelengths and with which he can probe into the center of a region forming a new star, and find out how stars form.

You once described Venus as a hellhole, hot enough to melt lead, with clouds made of sulphuric acid droplets. So that's unlikely to have any kind of life on it.

Jastrow: I think so. Except for one caveat about Venus. Early in its history studies of the evolution suggest it was quite pleasant for as much as a billion years or two. So life could have evolved and may still have remains of itself left somewhere. And it doesn't matter if the life is fossilized or still extant. If we find signs of any life on another planet in the solar system, we immediately know the answer to the number one question. Because, if life were extremely unlikely, it would occur only here and there in widely scattered solar systems. If you find it on two planets in one solar system, you know the a priori possibility is not one in a trillion trillion.

Have black holes been discovered?

Jastrow: My colleagues think they've discovered them. It's difficult to explain the properties of some stars. Townes and his colleagues have investigated the evidence for a black hole at the center of our galaxy. And there's good evidence that quasars are ordinary galaxies with massive black holes in the center.

What do you think of John Bell's theory that separate parts of the universe may be connected at a fundamental level and that once connected remain attached over distance by some unknown force? Apparently, paired particles moving away from each other seem to communicate at the same time. The theory developed from Einstein trying to work out why quantum theory was inadequate. These particles, though separated in space, twist at virtually the same time as if communicating with each other at faster than the speed of light. Is this plausible?

Jastrow: I learned from my friend Allan Sandage recently that general relativity did not forbid travel or communication in excess of the speed of light. Special relativity does. But I'm conventional in my thinking about it, and think that for whatever reason that limit is firm.

Is it generally believed that the universe will expand forever because there isn't enough dark matter [matter assumed to exist by some but not yet discovered] to stop it from expanding?

Jastrow: The available evidence suggests that. It's always possible more dark matter will turn up. My colleagues keep looking and every once in a while find another candidate. But so far that's the way it looks.

Have quasars been photographed 15 billion light years away from us?

Jastrow: If the size of the universe is 15 billion light years, then that's a little bit too far out. They've photographed the largest redshift corresponding to going more than half way out. I would buy 10. I'm not sure about 15.

Distinguished scientists say Star Wars was an absolute fraud. That it was simply impossible.

Jastrow: I don't agree with them. I think they're dominated by a kind of ideological thinking that stems from their position in the physics community on deterrence by the thinking of the ABM treaty rather than scientific deterrent.

Did you read the accounts saying it was really Edward Teller overselling it to President Reagan?

Jastrow: I've seen things like that. I form my own judgment on the technical merits, and I don't agree with those critics.

Have you had any ESP experience?

Jastrow: My mother, Marie Jastrow, had one which she put in a little book of hers, *A Time to Remember*, published by W.W. Norton (in 1979). And it is a psychic event.

[Mrs. Jastrow tells how, one bitterly cold night in the winter of 1921, her family was awoken by the sound of breaking glass. They found four glasses in their cabinet had cracked. Her mother said, "Somebody died. This is a sign. Someone near us died. Your glasses don't crack without reason while you're sleeping." Her father said, "Nonsense." But her mother, who was crying, explained why. At dinner one night when she was fifteen the family heard a crash in the bedroom and found a heavy mirror had fallen, though the hook that had supported it remained in the wall. A few days later they had news that the maternal grandfather was dead. Several days after the mysterious cracking of the glasses in the cabinet, Mrs. Jastrow's father found a black-edged letter in the mail. Her mother fainted. When she recovered she learned that her 22-year-old sister had died in Hungary of a throat infection. The letter was dated three days after the glasses had cracked. She cried for several days until one night she told the family she had dreamed of her sister who said, "Leave me alone! Let me go, already!" So she stopped crying, explaining that she understood the dream message: "You are not supposed to disturb the dead with your tears and lamentations. It is written. I will let my sister rest."]

What is your attitude to those experiences?

Jastrow: Anything my mother says is reliable. She's a very sober person. I take it at face value. That's why I keep an open mind on these things.

Do you think it's going too far to suggest that parapsychologists are studying some of the unusual events recorded in the Bible: changing water into wine could be called psychokinesis; prophetic dreams that come true, could be called precognition and so on. People have spoken of such things from early times and they seem to occur in every civilization.

Jastrow: They belong in a single class of events, I think that's the thought you are expressing, a class of events that are not explained by accepted means of communication.

In a recent book, Francis Crick suggested there's no such thing as soul or spirit or mind apart from the brain. That thinking and feeling are simply nervous reactions, the firing of neurons in the brain.

Jastrow: That happens to be my own view, but it's one that one, of course, can't prove.

But I thought that because of your mother's strange experience you wouldn't agree with Crick. That it's inexplicable.

Jastrow: On the face of it, but nonetheless, I have, as the result of many years of study, come to this completely materialistic or reductionist philosophy that Crick espouses. I'd like to believe otherwise, but at the moment that's where I am. I'm stuck in the middle.

Yet Sir John Eccles, the neuroscientist (and Nobel prize winner), believed in extrasensory aspects of the mind.

Jastrow: Yes, I know. Now he is able to maintain a dualistic philosophy without any problem. I've arrived at this reductionist conclusion just by reading books on the brain.

Do you have recurring dreams?

Jastrow: I dream about my parents a little bit.

Any interest or experience in hypnosis?

Jastrow: No I haven't.

Doesn't it intrigue you as an aspect of the brain?

Jastrow: I'm not sure I can cram any more experiences in my head.

What do you read?

Jastrow: Three newspapers every morning, New York Times, L.A. Times, and the Wall Street Journal. My colleagues go through the Washington Post and Washington Times for me, and cull material they know I'll be interested in. I read about 50 magazines: science and nature and astronomy journals. I haven't been reading the physics journals for quite a while. Then I read The New Republic, and Foreign Policy Review. I get through a lot of reading, skip reading. I haven't done much reading of books lately. There's a pile next to my bed I want to read.

Did you read novels in your young days? H.G. Wells?

Jastrow (chuckled): I read all that stuff when I was younger. But I haven't been reading much fiction lately.

Are you the only scientist in your family?

Jastrow: Yes.

What did your father do?

Jastrow: He sold cars.

Do you have a family of your own?

Jastrow: No. I'm single and childless.

Did anyone direct you toward success?

Jastrow: No, I was really inner driven.

I noticed in the early days at Mount Wilson there was an armed guard outside the observatory because of roaming mountain lions.

Jastrow: We have a bear problem now and then, but I don't think mountain lions have been sighted recently.

To Jastrow's delight, in February 1994, NASA announced plans to launch two small unpiloted flights to Mars in November 1996, to orbit Mars and conduct a two-year photographic and "remote-sensing" survey of the planet's weather and geology. This will be followed by a decade of observations from spacecraft as well as landings on the planet. Scientists expect to gather information about its annual climate cycles, surface mineralogy and chemistry, and to pinpoint suitable landing sites for manned spacecraft. Early in the next century they hope to bring back soil and rock samples.

On April 7, 1994 I spoke with Robert Jastrow again.

You seem to have outdone H.G. Wells in your speculations about the human brain in *The Enchanted Loom.*

Jastrow: It will be more of an adjunct to our intelligence in the shape of a partner, like Data in "Star Trek," or that intelligent creature in "Star Wars." A somewhat narrow but very strong intellect. We will create an intelligence of human-like capabilities, perhaps not human wisdom, that we will cause to evolve rapidly because of its tremendous value in enhancing the productivity of our lives. And in the long run, if you look beyond the time scale of the next few thousand years—this device is eventually certain to outstrip us, because it can evolve without limit. At first it will be an electronic slave.

You write that at first it will be an electronic slave, then "I'm sorry to say it may actually become our master."

Jastrow: But not in a way that we would ever know. The oysters on the seashore are not aware that they've been overtaken and surpassed.

Will this new intelligence be physically attached to our brains?

Jastrow: I don't think there will be anything we plug into the interface between the biological form of intelligence and the silicon-based intelligence.

People won't be born with it?

Jastrow: No. My thought is that homo sapiens has not changed for 50,000 years, so we are a finished story. And the history of life tells us that out of the highest forms on the planet at any time there always evolves a newer and higher form yet. So there's no reason why that many-million-year history should be interrupted at this particular level of intelligence.

You think intelligent life in other galaxies may already have that higher intelligence?

Jastrow: Think of the time scale. On the average, the planets in other galaxies are 3 billion years older than our planet earth. Because the universe is 15 billion years old and the average age is half of that or $7\frac{1}{2}$ billion, and that's 3 billion more than $4\frac{1}{2}$ billion. So you ask yourself what will develop in the course of 3 billion years? Then you see in the next million years or so, judging by the past history of life on this planet, we are going to see the emergence of a form which is superior to homo sapiens—as much as we are to the 3-million-year-old ancestors of mankind, the transition between the great apes and the genus homo. They lived in Africa 3 million years ago, and had primitive tool culture, and a smaller brain and body than ours. The body of a 12-year-old child.

But you say creatures in outer space are more likely to be disembodied, what some people call spirits. Maybe, you suggest, they can materialize and dematerialize.

Jastrow: Yes. I ask you, again, to think what a billion years is. It's a thousand times a million.

You have said that you are sure such creatures have magic powers by our standards, but you don't know whether we'd recognize them if we saw them!

Jastrow: The person who is thoroughly sympathetic to this line of thinking is Marvin Minsky, who's actually in the field [artificial intelligence].

If they're disembodied could they survive the destruction of the universe?

Jastrow: That's an iffy question. The destruction of the universe, you mean, in a forthcoming collapse? No, I don't see how, if that is indeed our fate, any organized form of matter or energy could withstand it.

Being nonmaterial doesn't save them?

Jastrow: No. And, by the way, who anticipated me in this many years ago, was Bernal, the British biochemist, in *The World, the Flesh, and The Devil*, an extraordinary, prescient book, written in the 1920s. He anticipated everything that's happening.

You must be pleased they've decided to return to Mars in 1996.

Jastrow: Yes I am. In fact, I was just writing a letter to the head of NASA urging that the matter of Martian life be reopened. The test for microbes on Mars satisfied all the checks and cross-checks, and was itself positive evidence, but no one could find any organic matter, any of the building blocks of life on the surface, though they looked with very sensitive instruments. So they said if there's not organic matter, no life. New evidence has come up suggesting the negative findings were premature. A student at MIT ran a test on a sample of Antarctic soil with a sparse population of microbes probably. By the standard lab methods he detected .03% organic matter. By the

Viking test he could find nothing. Which means that the Viking test is 100,000 times less sensitive than everyone had assumed. So there could have been a rich population of microbes in that soil and it would not have seen them. This instrument's results are the main reason for the negative findings on Martian life. I'm glad we're going back and I'm sure in the next century we'll have, if not people, automatons, intelligent silicon-based geologists and biologists looking for life on Mars. And I would guess they'll find it.

You mean those sophisticated computers will do the looking?

Jastrow: Yes. You shouldn't really call them computers. Think of Data in "Star Trek." The only thing they'll lack is the package of emotions that provide survival value.

Would it be insulting to call them computers?

Jastrow: No. But it misleads. A computer is only as good as the programs you put into it.

That was Arno Penzias's objection to thinking computers could do such things as you mention. But you mean something more sophisticated.

Jastrow: Something evolving. Remember the complexity of the human brain depends very much on the number of interconnections. The neurons. We have 10 billion neurons, each one a little computer in itself.

Will they have initiative?

Jastrow: Yes. Initiative means a wish to perform something you've been programmed, you've been motivated for genetically, or by training.

Because we've been programmed and motivated, too, is that it?

Jastrow: Yes. We've been programmed by our genetic inheritance through our parents.

I spoke with Robert Jastrow, again, in December 1994 having learned that a new sighting in the sky made the observers believe the universe is much younger than Jastrow believes. "The issue," he said, "is entirely whether or not they are correct in believing the object they were observing is the center of a coma cluster, or an outlier. They mentioned the distance to the center of this cluster and the distance was shorter than others might expect who say the universe is 20 billion years old—implying a younger universe that hasn't had as much time to expand as far."

When, a few days later, the latest photographs from the Hubble Space telescope gave a clearer picture of our expanding universe, Jastrow advised me to check with the expert on the subject, his astronomer friend Allan Sandage. [See Chapter 24 for Sandage's views.]

Chapter Ten.　Charles Townes

"Most of my successes have come out of failures."
—Charles Townes

When I spoke with the 79-year-old Nobel Laureate Charles Townes, he was about to take a high-altitude flight to spend the night studying the stars over California. He's often aimed high, climbing the Matterhorn when he was 40.

As a boy, Townes lived on the wooded outskirts of Greenville, South Carolina, with two brothers and three sisters. He was fascinated by the birds and insects that shared the land with them. At Furman College he majored in modern languages and physics, filling free time as curator of the college museum, member of the band, glee club, swimming team, and newspaper staff. He left college for Duke University in 1935 as the best scholar in the senior class and with a medal for outstanding work in the sciences.

Physics, "because of its beautiful logic," became his passion at Duke, where he also studied French, Russian, and Italian. In 1937, with an M.A. degree, he went west to Caltech, where he was awarded his Ph.D. two years later. During World War II at Bell Labs, Townes helped to design radar systems to enable aircrews to navigate and bomb through clouds and at night. He also studied microwave spectroscopy and discovered high-resolution microwave spectroscopy of gases. Microwave spectroscopy uses radar-like radio waves to examine the structure of materials. In 1952 he headed Columbia University's physics department. There, his molecular wave experiments led to his discovery of the "maser" (microwave amplification by stimulated emission of radiation). The maser amplifier has proved to be of great value for radio astronomy, because it detects weak signals.

In 1958, with Navy scientists, he used a maser-equipped radio telescope to check the temperature of Venus. His maser also made an atomic clock possible without electric circuits. If run for three hundred years, it will keep time to within one second. It tested Einstein's special relativity by measuring the frequency of an electromagnetic wave traveling in the same and in the opposite direction to the earth's motion as it orbits around the sun. After the most precise physical experiment ever made—with an accuracy of one part in a million million—scientists established in 1959 that Einstein was right: the velocity of light is a constant 186,000 feet per second regardless of the movement of the observer.

With his brother-in-law, Arthur Schawlow, Townes developed the "laser" (light amplification by stimulated emission of radiation)— a concentrated beam of light capable of bloodless surgery and of improving radio and TV transmissions. It can burn a hole through steel. It has also made possible supermarket scanners, compact discs, and laser art.

He believes that the emotional experience of making a scientific discovery is similar to what some people describe as a religious experience or revelation.

In 1961 he became provost of M.I.T., where he also gave an occasional lecture and continued "limping along" with his researches. Three years later his laser won him the Nobel prize in physics.

Townes met his wife, Frances Hildreth Brown, while skiing, and they have four daughters.

He is considered something of a daredevil, never able to resist a challenge according to his brother-in-law, Arthur Schawlow.

Always on the lookout for fresh fields to conquer, he's raised his sights to way above the Matterhorn. After our conversation on February 14, 1994, Townes hoped to solve a cluster of cosmic puzzles: Why were a group of stars behaving so mysteriously? Were they in danger of being swallowed by a nearby black hole? Was it really a black hole or some as yet undiscovered phenomenon?

Townes: I'll be flying tonight in a C 131 plane run by NASA. It has a 36-inch telescope in it and we go up above the atmosphere to about 40 to 45 thousand feet so we can observe in the red wavelengths which don't penetrate the atmosphere. We slide back the door in the fuselage and operate the telescope all night. I frequently do this. We're after a number of goals. We're studying the center of our galaxy in the constellation of Sagittarius, the Orion nebulae. There are many unusual things going on. There seems to be a black hole there, but it's not behaving the way everybody expects a black hole to behave. It's not putting out as much energy. So we're trying to understand that. Clearly things are falling into it. Also there are some very unusual stars there. One doesn't find a group of stars like that anywhere else. So that's another puzzle.

Is it definitely established as a black hole?

Townes: The evidence clearly shows a strong concentration of mass and we don't know anything it could be except a black hole.

How did you discover the laser?

Townes: It was a bright, beautiful day, and I had slept in the same hotel (in Washington, D.C.) all night, with my brother-in-law, Art Schawlow. And I simply woke up and I didn't want to wake him. I sometimes say it was because of my children, because they made me accustomed to waking up early in the morning. So I walked out to the park and sat down to think of the coming day's meeting, which was to discuss how to generate high-frequency waves. And that's when the idea occurred.

How do the laser and maser differ?

Townes: A laser is a maser operating at optical wavelengths. The first system of its type was a maser operating at microwave wavelengths. (The maser is an electronic device that absorbs microwaves, infrared rays etc, amplifies them and emits them in a narrow, very intense beam.) It also produces very uniform radiation, generally not highly intense. Now there are intense masers in space, more intense than the total energy of the sun.

When it's said a laser beam is hotter than the sun, what area of the sun?

Townes: The surface or the center. It can be fantastically hot.

Other than medical, industrial, and communication uses of the laser, amazing in themselves, is there anything else the laser could be used for?

Townes: It's enormously useful in science. It has many, many scientific applications. Many labs use it. It has improved scientific instrumentation very importantly. The laser is a kind of marriage between electronics and optics. And it touches almost any technical field. We also have laser art and laser shows.

Has it lived up to expectation in the medical field?

Townes: I would say more than that. There continue to be further and further medical applications of it.

I'd heard it was at first said to be painless for eye surgery but that wasn't always so.

Townes: Oh no. It's less painful. It's effective in cancer treatment, but it doesn't cure all kinds of cancers. For certain skin cancers it's very effective. But it's difficult to get it inside the body. However a lot of people are very clever with getting it inside the body, with fibers and so on. What might develop importantly is the use of lasers in a certain spectral way. Cancer absorbs certain chemicals which react to light of certain wavelengths. Then the laser looks in these wavelengths and the cancer cells are destroyed and the non-cancer cells are not. That has been studied and experimented with. It's been somewhat successful in the laboratory. It's promising. If the cancer is difficult to pick out and the cancer cells are mixed with healthy ones this would be a very good way of eliminating them.

Are you continuing your research on lasers?

Townes: I don't generally do research on lasers themselves. I use lasers in my research in astronomy and astrophysics. In the equipment we fly at high altitudes we're using three lasers to control the equipment and properly adjust it. I'm also using it in another astrophysical experiment to measure the sizes of stars. There one needs very careful control of distances. And the laser is just perfect for that. Laser interferometry (measuring wavelengths of light and analyzing small parts of the spectrum) is now very well developed and used by many people. I'm also using it to detect infrared radiation in that same experiment.

Is the idea of the laser as a death ray just fantasy?

Townes: It's not a fantasy. It can kill people. But it's not a good weapon of destruction.

Even though it's hotter than the sun?

Townes: That's correct. A bullet is much cheaper and more effective generally. The laser is important in guiding the aiming of a tank or guiding the aiming of a missile or bomb dropped from a plane. It allows the missiles to hit the target accurately rather than scatter them.

Used in Desert Storm I presume.

Townes: It was and was very effective there.

How effective?

Townes: A laser can guide something right on target, to within a foot.

So there should have been no misses.

Townes: Unless something went wrong. And things do go wrong.

Did you think SDI (Strategic Defense Initiative also known as Star Wars) was plausible?

Townes: No. I never felt that SDI would achieve the goal Reagan set for it, and a good fraction of the money was somewhat wasted. That doesn't mean all the work was useless. I'm very much in favor of thinking about it carefully and investigating it. But I'm pretty convinced it won't work. I'm not against somebody else keeping working at it. I might look at it again myself.

You know Bell's theory and the possibility of something moving faster than the speed of light . . .

Townes: That does not allow [for] things faster than the speed of light. It allows [for] things which when casually interpreted might seem they're faster. But they never really are, in the sense that you cannot communicate or send energy faster than the speed of light. I'm convinced that's true. But I'm always ready to be disproved. It would be very interesting if there were something, but I believe there's nothing. That doesn't mean you can't do experiments which give an impression there's something faster than light. But it is never real communication or transfer of energy. I believe in Bell's theorem. We've done experiments on it. It's perfectly correct. But it doesn't say you can send energy faster than the speed of light.

I thought the two particles seemed to be communicating instantly.

Townes: You say "seemed to be." That's correct. But they aren't really. It's a condition in quantum mechanics set up beforehand which produces that. It's not direct communication, or any flow of energy between them. But the appearance is there.*

Is it worth more experiments?

Townes: Oh yes. There are many aspects of quantum mechanics which are counterintuitive and people have not taken them seriously until the last decade or so. [See conversation with Arthur Schawlow on Bell's theory in Chapter 11.]

Do we absolutely know the nature of light?

Townes: When you say absolutely I would say of course not. We never pretend to know anything absolutely. If you mean, do we think we understand light pretty thoroughly, I would say yes.

Even the peculiar behavior of light as a wave and a particle?

Townes: We think we understand it. That doesn't mean we can be certain. There may be some surprises down the road.

Even the double split experiment?

---

* In some atomic processes a pair of photons (wave-particles of light) behave as if still connected, no matter what distance separates them. In 1964, John Bell, a physicist at CERN, provided mathematical proof that if measurements on both particles had a greater degree of correlation than would be expected by chance, it didn't mean they were communicating at speeds greater than that of light. He concluded that either quantum theory was flawed, that objective reality did not exist in the subatomic world, or that all parts of the universe were infinitely interconnected.

Townes: Oh yes, we understand that. We can work it out mathematically and show just exactly what it should do. We can predict that. If we can predict it, you might say we understand it. Now understanding is a kind of vague term, and because some of it is counterintuitive you might say we don't understand it. But if we can predict what it should do and do an experiment, and say yes, that's just the way it behaves, I would generally call that understanding.

Do you believe in God?

Townes: Yes.

Very few physicists do.

Townes: Relatively few. But a surprising number actually, and it's becoming somewhat larger. The interaction between science and religion has increased, I think, in the last decade or so.

Do you believe purely on faith?

Townes: I would say I feel it intuitively. I think my prayers have been answered. On the other hand to prove that scientifically is somewhat like the problem of telepathy. It's my own judgment over my experience that makes me believe in God.

In a recent book, Francis Crick writes that there is no spirit, no mind, no inner man, that all thinking is simply the firing of millions of neurons in the brain.

Townes: Would I dispute him? In the first place, I'd say we don't know. I would say, secondly, it may well be that. We don't understand the brain and how we come by our thoughts. It's so complex one could raise the question whether a complex system can ever understand itself. Certainly a higher order system can understand a lower order system frequently. It's clear there are a lot of things about the brain we don't know. Also, do any of us have any freedom in what we say and think? We pretend we do. We think intuitively we do. We feel we do. But is it real? Scientifically there's no defense for saying it's real. It's all mechanistic, hence we have no choice at all. We have no free will. I would prefer to err in the direction of thinking what my intuition says, that yes I do have some kind of choices open to me.

Do you believe in an afterlife?

Townes: I believe in the possibility. Whether it's real or not I don't know.

Do you believe Jesus Christ was God?

Townes: That he was part of God I could say, yes in a sense he was, and so are you. Christ was closer to being God-like than most of the rest of us certainly.

What do you think about intelligent life on other planets?

Townes: The probability is very small. That doesn't mean it's zero. It's possible we're the only ones here. I feel sure that life is very rare.

You don't believe like Jastrow that there's primitive life on Mars, microbes?

Townes: There might be. There's no evidence that there is.

He thinks there is evidence.

Townes: I don't agree.

Any interest in UFOs?

Townes: In a casual way, but I don't take them very seriously.

How many light years away do you think the edge of the universe is?

Townes: The universe has a size of about 15 billion light years.

PERSONALITIES

Kip Thorne says that in our time Feynman was the only scientist who was intuitive in the way Einstein was. Do you agree?

Townes: No. He had some qualities Einstein did, but other scientists do, too. Intuition is not uncommon among many of the better scientists.

What did you think of Feynman?

Townes: Always enjoyed him. Quite a character. (He chuckled.) Definitely very bright. I admired him in many ways.

Was the title of his biography, *Genius*, justified?

Townes: In terms of what we generally mean by genius, yes, sure.

You met Einstein.

Townes: Yes. A friend at Princeton said "Would you like to meet Einstein while you're here?" and I said, "Sure." And we had a good long chat at Einstein's home. I found him a very modest and nice person and interested. What surprised me a little bit was that I was working on some fairly technical things for Bell Labs and he was very interested. But after I thought about it, I said, "After all, he's been involved with patents." I found him very low key, a pleasant, interesting person. Very different from Oppenheimer who was very fast and a bit of a showoff. I knew Oppenheimer very well.

Unjustifiably treated?

Townes: Unfairly treated. To some extent it was his own personality kind of asking for that. But it was still unfair. He was fantastically quick and he just seemed to understand everything. Very quick and very good memory. Anybody who spoke to him immediately could see that guy's bright. Very different from Bohr in that respect. But Oppenheimer, in the long run, did not do as deep work as Bohr.

Was Bohr difficult to understand?

Townes: A little cryptic and he didn't know how to express himself very fluently. He would express things which were not clearly stated and maybe not even right.

Which contemporary scientists most impressed you?

Townes: Einstein was exceptional. I think he's the one I'd pick out.

I.F. Stone thought Einstein was an almost God-like presence. Did you feel that?

Townes: No.

Do you think alone?

Townes: Generally, But I also discuss things with people. I frequently tell my students or friends that I get a lot of ideas from discussions. I don't purposely say, "Now I'm going to think with somebody," but I find discussions quite important.

What do you still want to do?

Townes: I'm working on two projects. One is with the airplane, which I feel is over the hump and doing very well. Other people are getting into that field. Characteristically I try to explore new fields, and when I sense it's going well, I leave it. I'm not needed any more. Right now a field that I have studied more recently, is interferometry: to do very high angular resolution on astronomical objects. It's a difficult field but one I felt was ripe and needs to be developed. It's been carried on, off and on, since Michelson's day. People try it and drop it because it's too difficult. I think now the technology is here and I'm working at it with some success. I hope to see the field well established and going well. So that's the immediate thing I'd like to do and see done. When it's going well I can feel happy in giving it up.

Can you tell the layman what that would achieve if successful?

Townes: It's essentially like getting a microscope on the sky. Objects we are looking at are so far away we don't see many details. And if we could have a microscope, you'd see a lot more details.

PERSONAL

What was your father?

Townes: A lawyer interested in natural history as I was. He might have been a scientist if he'd come along in a later generation and a different part of the country. (He was from South Carolina.) I had an older brother who was a biologist. He's dead now. He worked at the University of Michigan and set up an entomological institute down in Florida.

You're reputed to have been a daredevil in your youth, climbing the Matterhorn for example.

Townes: That was not exactly my youth. (He laughed.) I was 40. I wouldn't call myself a daredevil. I'm just curious. I like to try new things and I'm willing to take a few chances. I'm not a daredevil in being careless. I was careful to get a good Swiss guide in 1955.

A tough climb?

Townes: Oh yes. It's a long climb and there's lots of exposure. That's what bothers people. Going from 5,000 feet to where any misstep could really take you down.

What else have you done?

Townes: First time I ever went up in an airplane, in Southern California, I flew it. It was advertised that anybody who could drive a car could fly a plane. I could drive a car. I got together with another student and we had somebody sitting beside us telling us what to do. We took off and we landed. I wouldn't consider that enormously dangerous. We had somebody there who if we made a mistake could grab the controls. I've done some solo flying in California, but my wife got worried so I decided if I had to get somewhere fast I'd hire a pilot.

Have you had any major disappointments?

Townes: I've had all kinds. I sometimes say most of my successes have come out of failures. A failure makes you try to do something else. For example, I came from a small school in South Carolina and I couldn't get support to go to a good graduate school. The only place I could get support was Duke University which was nearby, and I got a teaching assistantship there for a year. I applied to all kinds of other places and was turned down. Then I tried for Duke's one fellowship and didn't get it. I decided, gee, I'm just going to break loose and go to the best place I know. And I'd saved $500 and thought I'd see how long I could last on that. So I went out to Caltech. So that was a failure. But I went to Caltech. That sort of thing.

Like one door closes, and another opens.

Townes: That's right. You find something else that's more likely and you change track. So I didn't stay in Duke, which was a good university, but Caltech was much better.

Ever been hypnotized?

Townes: People have tried to hypnotize me but it didn't take.

You were at Duke the same time as the parapsychologist J.B. Rhine who was investigating the extrasensory world. What was your impression of him?

Townes: I felt he made a serious effort to examine the questions. And I felt it was very important to examine them.

Do you think telepathy is possible?

Townes: Telepathy and transformation of information between people in a way we don't realize is possible. But it may be the slight twitch of a finger.

I mean with people thousands of miles apart.

Townes: I see. I was just explaining my caution about the word. If you mean some new mechanism of transfer of information, it is possible. Anything's possible. I doubt if it's real. But I think there are reasons for experimenting with it, and I'd be fascinated if anything turns up.

You've had no such experience?

Townes: No.

No dreams that came true?

Townes: Oh, sure. I've had dreams come true. And I've had things that seemed peculiar coincidences.

How do you explain it?

Townes: I dream about things I'm thinking about, and sometimes they do happen. It's a question of coincidence. My dreams are closely connected with thoughts, and ideas, opinions, and work.

What are your politics?

Townes: I'm a middle of the roader. For presidents I've generally voted Democratic. But I've been skeptical of some of them.

What do you think of Clinton?

Townes: I wish he were better. I think he may be fairly good. He's made some mistakes, but he's done some pretty good things. I'm hopeful. Just waiting to see whether it will work out or not.

Who owns the patents to your discoveries?

Townes: I gave the maser patent to Research Corporation, although it's in my name. And the maser patent covers both masers and lasers. So any laser made and sold would have to get a license from Research Corporation. Now the laser was patented by Bell Telephone Lab in the name of Schawlow and myself, but it was the property of AT&T.

So anybody wanting to make a laser would also have to get a license from AT&T.

Is Research Corporation a private company?

Townes: It's a public corporation dedicated to supporting research in universities, set up by Cattrell, another inventor. He made a good deal of money with his inventions in industry and then set this up as a way of helping university research. It's a public charity corporation.

If you were starting over would you go in the same field?

Townes: I might, but I would guess more likely biology, perhaps molecular. I was always interested in biology, but it has become particularly interesting now—more fundamental. They're finding out a lot of interesting things.

After speaking with Townes I read in a decade-old news clip that he had not only headed a study of the MX missile, but also as a member of the Pontifical Academy of Sciences in Rome, advises the pope on scientific matters. Interviewed by science writer, Richard F. Harris, Townes named stubbornness as one explanation for his success. "When I was working on this maser," he told Harris, "I had two Nobel laureates come into my office and say, 'You really have to stop this nonsense. You're wasting not just your time, you're wasting money ... and we know it's not going to work, you know it's not going to work, so you just ought to stop. About three months later we had it working. Stubbornness must, however, be tempered with objectivity ... You must also be willing to make some mistakes."[2]

Chapter Eleven.
Arthur Schawlow

"I'm fascinated by the nonclassical nature of light and the
nonlocal nature of quantum mechanics"
—Arthur Schawlow
"Astronomers are very brave and bold, and make vast assumptions
based on very little data."
—Arthur Schawlow

Most scientists I questioned were atheists or agnostics. John
Wheeler was, at best, on the fence. So I was interested to learn that
Nobel laureate Arthur Schawlow's favorite book is not *The Origin of
the Species* or the collected works of Isaac Newton, but the Bible. This
he told Carl Irving, when interviewed for *The San Francisco Examiner*
in 1985. When I spoke with Schawlow almost a decade later, he not
only confirmed his faith but said his brother-in-law, Charles Townes,
is also religious. Townes is someone Schawlow clearly admires.
"Charlie's a man who can't resist a challenge," he said. "At about 40
he decided if he was going to climb the Matterhorn he'd have to do
it soon. So he went into training in Austria, and climbed some lesser
peaks and then went ahead and did it." Schawlow's challenge has
been to help his autistic son, Artie. When he got the 1981 Nobel prize
for research on laser spectroscopy, he used some of the money to buy

a machine to communicate with him. On the Stanford faculty since 1961, Schawlow taught freshman physics to classes often topping 200, interviewing at length those who planned a medical career. He found this time-consuming, but said it sharpened his wits. Besides Artie, Schawlow has two daughters, Helen and Edith. Both have college degrees. His wife died in 1991.

I asked Schawlow: How do you try to communicate with your son, Artie?

Schawlow: When I went to Stockholm in 1981 for the Nobel prize, I met the mother of an autistic girl who had written a book about her. The mother had a Ph.D in psychology she told me about a young man who couldn't talk, but could type things on a hand-held device like a calculator that printed on paper tape. She asked him, "May I have one of your tapes?" And he replied, "No." She said, "Why not?" And he said, "You can't read it when the sun shines." It was thermal printing and fades in the sunlight. And I thought, "Oh my goodness, if my son understood things like that he'd have no way of telling me." So when I came back I found out what this device was: a Canon communicator. A flop at first. He'd just type garbage: xxxxx, zzzzz, that sort of thing. So we went through a number of things like trying to identify letters with touch and tell. He could do that, and match words to pictures. And then I programmed the first laptop computer, the Epson HX20, so nothing would happen unless he hit the right key. It would show a word with a dash under every letter. And when he finished the word it would print it out. He liked that and kept on until he'd used up all the paper, and stuffed it in his pockets. But it's still not communicating.

Then we learned about communication boards where you can point to words on a board and can choose things, like what kind of snack you wanted by pointing to it. At one point I asked him to confirm his choice by typing it on the communicator. He was willing to do that. Shortly after, when he was 27, we were in a Baskin-Robbins ice cream store. And he'd had his ice cream and he waved his arms and you could tell the way he was waving them he wanted something in that direction. So I acted dumb and said, "Come into the car and we'll get the communicator and you can tell us," not

Continues on page 239

1. LINUS PAULING speaking at Caltech in 1986. Fellow Nobelist James Watson called Pauling's lecture style "dazzling. There was no one like Linus in the whole world. The combination of his prodigious mind and his infectious grin were unbeatable."

—Courtesy of Caltech.

2. ATOMIC MODEL: In 1963, the year he won the Nobel Peace Prize, Pauling displays his model of the alpha helix. The previous year Watson and Crick won the Nobel Prize for their discovery of the double helix—the heredity molecule. It was based on Pauling's work on helical structures. In fact, he nearly beat them to it.

—Courtesy of Caltech.

3. THE TEAM: Pauling and wife, Ava Helen, holding hands in December 1966. He called her his "second conscience." That year Pauling said he hoped to live another 20 years. He lived for another 27, to age 92.

—Courtesy of Linus Pauling.

4. CALTECH VISIT in 1974 found Pauling in a happy mood. He is still the only one in the world to win two unshared Nobel Prizes—one for chemistry and one for peace.

—Courtesy of Caltech.

5. ON CAMPUS at Caltech during his 1974 visit Pauling exchanges ideas with students. "I think about a problem all the time," he said. "Wherever I am: in bed, going for a walk, traveling."

—Courtesy of Caltech.

6. LAUGHTER came easily to Pauling, who had a lively sense of humor. Here he is in 1982, still happily at work. He worked up to a few weeks before his death in 1993.

—Courtesy of Caltech.

7. RICHARD FEYNMAN enchants a group of students at Caltech in 1965.

—Courtesy of Caltech.

8. FEYNMAN'S PEERS: Inevitably Feynman is the most informally dressed for a Nobel Prize Winners Dinner in 1970. From left: Carl Anderson, Murray Gell-Mann, Max Debruck, Richard Feynman, George Beadle.

—Courtesy of Caltech.

9. SPELLBINDER: Feynman holds his audience's attention during a Caltech seminar in 1978. His lectures were often punctuated by laughter and bursts of applause.

—Courtesy of Caltech.

10. PAUL DIRAC had a reputation among friends and colleagues as a man of long silences and few words who hated redundancies of any kind. He was once handed a manuscript which had written on it, "Must not be published in any form," and he erased the words "in any form." With age he mellowed and became more willing to talk.

—Courtesy of Florida State University.

11. VICTOR WEISSKOPF when director of CERN in Geneva, Switzerland. "I'm interested in everything . . . Yesterday I was at a concert—the Beethoven sonatas. That such things were made by humans is something which . . . makes life worth living."

—Courtesy of Photo CERN.

12. HANS BETHE at Cornell University. He headed the Manhattan Project's theoretical division at Los Alamos. Bethe says it is possible to travel to Mars, Jupiter, and Pluto, on a spaceship powered by 2,000 atomic bombs, but the problem would be that "nobody likes 2,000 atomic bombs exploded in the atmosphere."

—Courtesy Cornell University.

13. JOHN WHEELER knew Einstein, Bohr, and Werner Heisenberg. "We go down and down from crystal to molecule, from molecule to atom, from atom to nucleus, from nucleus to particle, and there's still something beyond both geometry and particle. In the end we have to come back to mind."

—Courtesy of University of Texas, Austin.

14. GEORGE WALD: From work on the eye to the nature of consciousness in the universe. "A physicist is the atom's way of knowing about itself . . . This is a fantastic universe from the point of the extreme improbability of many of its absolutely basic properties."

—Courtesy of George Wald.

15. THE TEAM: Arno Penzias [left] and Robert Wilson who discovered cosmic radiation, persuasive evidence for the Big Bang, stand in front of the horn antenna of the radio telescope they used at Holmdel, New Jersey, to make the discovery.

—Courtesy of AT&T.

16. ARNO PENZIAS: "People are uncomfortable with the purposefully created world. To contradict purpose, people tend to speculate about things they haven't seen yet, like missing mass, which would allow the world to collapse back on itself."

—Courtesy of AT&T.

17. CHARLES TOWNES in his office at the University of California at Berkeley in 1988. He came up with the idea of the maser, three months after two Nobel laureates told him he was wasting time and money by working on the project. Now, a professor of physics, he often takes high-altitude night flights on voyages of exploration.

—Courtesy of Charles Townes.

18. ROBERT JASTROW, director of Mount Wilson Observatory in Pasadena, California, holds a transparent globe of the constellations. "There are a lot of stars and planets out there like ours . . . I personally am interested, not only in physics, but in life in the cosmos."

—Courtesy of Robert Jastrow.

19. ASTROPHYSICIST: Robert Jastrow: "We've been on this plant for about 100,000 years as the defined species and the average beings around us are billions of years younger or older."

—Courtesy of Robert Jastrow.

1

2

4

5

6

7

8

9

10

11

12

13

14

15

16

17

18

19

20

21

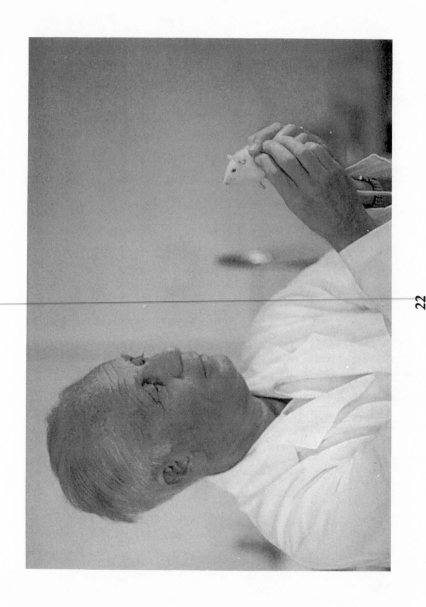

24

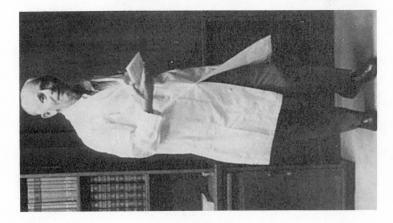

23

26

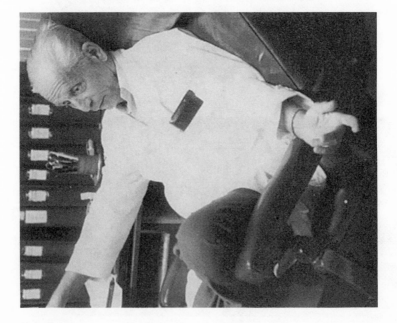

25

27

28

29

30

31

32

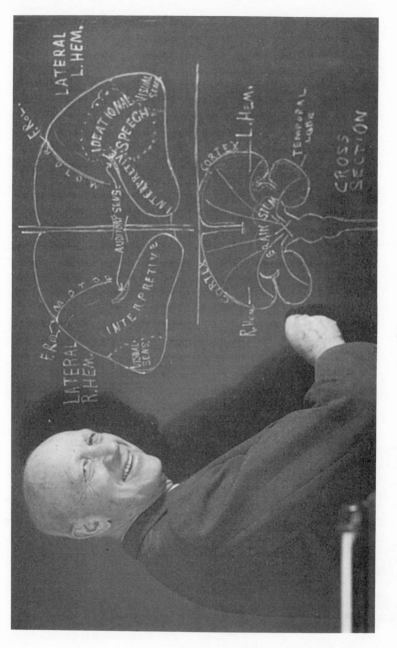

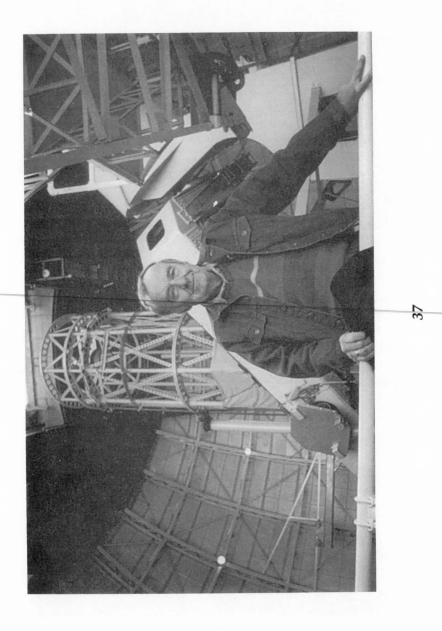

20. ARTHUR SCHAWLOW: With Townes he wrote the seminal paper on the laser or "optical maser." He also worked on how to foil counterfeiters.

—Courtesy of Arthur Schawlow.

21. HAROLD UREY, professor of chemistry at the University of California, San Diego, was persuaded to pose with the symbols of his success. "There are many stars in the universe that should have planets similar to ours. If we found a planet with the same or very similar conditions to earth, then it is probable that life would evolve there."

—Courtesy of University of California, San Diego.

22. HANS SELYE studies a white rat. "We have no other way of telling an animal is old but by looking at it."

—Courtesy of Countess von Schwerin.

23. SELYE enters his office in Montreal, Canada, for a conversation with the author. "Variety of experience is not only the spice of life but possibly the key to longer life."

—Courtesy of Heather Pratt.

24. STRESS EXPERT: "Whenever you are exposed to stress" says Hans Selye, "a number of biologic reactions go into your body and 99.9 percent of them are reversible. But there is one minute fraction that leads to irreversible results, what I call chemical scars."

—Courtesy of Heather Pratt.

25. THE SKEPTIC: Selye said : "I don't think it's biologically possible to love thy neighbor as you love yourself, because somebody commands it."

—Courtesy of Heather Pratt.

26. THE OPTIMIST: "Owing to evolution man has developed such perfection of the central nervous system that we can have purely reflex responses, or purely emotional responses—including cruelty—subjected to the control of logical analysis."

—Courtesy of Heather Pratt.

27. HENRI ELLENBERGER: Freud "suffered from heart ailments, depression, and from psychosomatic ailments. They were sometimes agonizing. He thought he was about to discover a great secret, perhaps by analyzing dreams."

—Courtesy of the Menninger Archives.

28. THEODOR REIK was considered by Freud to be one of his best students. "Freud said to us, his intimate circle of students, 'Never give advice to your patients except in the form of the ten commandments. Don't do that! Don't do that!' "

—Courtesy of Murray Sherman.

29. PSYCHOANALYST Murray Sherman, Reik's student, says, "Never say never!" but praises Reik's "gift for feeling and communicating the unconscious," adding, "Psychoanalysis is a process of discovery."

—Courtesy of Dorothy Gloster.

30. HARVARD PYCHOLOGIST Henry Murray who developed the Thematic Apperception Test [TAT] and used it during World War II to test wouldbe members of the OSS [forerunner of the CIA].

—Courtesy of Nina Murray.

31. HENRY MURRAY in his sixties as director of Harvard University's Psychological Clinic. "Jung would take people that were pretty damn psychotic on the insane side, but think they were treatable, and communicable, and relatively normal. Other practitioners wouldn't have anything to do with them"

—Courtesy of Nina Murray.

32. CORNELIA WILBUR: "It's been my own experience that multiple personalities are very definitely curable . . . I've fused several. One of them in about three years."

—Courtesy of the Chandler Medical Center, Lexington, Kentucky.

33. ASHLEY MONTAGU discussing world affairs and human nature with Albert Einstein in the latter's Princeton study. "I said, 'Professor Einstein, international law exists only in the textbooks of international law.' "

—Courtesy of Ashley Montagu.

34. WILDER PENFIELD made this diagram of the human brain simultaneously using both hands to draw the left and right halves—which explains the smile. "It would take a lifetime and a new order of experimentation and morality to completely recover the stream of memory."

—Courtesy of the Montreal Neurological Institute

35. ROGER SPERRY, professor of psychobiology at Caltech. His ideas about the nature of the brain, based on his experiments, overthrew the long entrenched behaviorists. He is shown illustrating his ideas. "I suspect that if mosquitoes, hummingbirds, and insects can be conscious as many people agree . . . the secret programming or processing is not all that complicated."

—Courtesy of Caltech.

36. TORSTEN WIESEL shakes hands with the king of Sweden during the Nobel prize awards of 1981. "I have concluded that a balance between the environment and heredity, nature and nurture, have an effect on the operations of the brain."

—Courtesy of Torsten Wiesel.

37. ALLAN SANDAGE in the mid 1980s with the Mt. Wilson Hooker 100-inch reflector as his companion. Sandage learned astronomy from Edwin Hubble, who discovered the universe is expanding. "The tremendous controversy [since 1994] is the scale of the universe."

—Courtesy of Allan Sandage.

thinking that he would. And then [after several years of typing out "garbage"] he typed out "shoes." There was a shoe store there. So we got him shoes. Two days later my wife took him out and he typed a lot of things. One was "I want to go home." He was staying at a hellhole of a state hospital, but there wasn't any other place to take him.

Is it still a hellhole?

Schawlow: Yes.

It's a question of money, is it?

Schawlow: No, there just aren't many places that can handle him. He's big, strong, and fast, and occasionally blows up. He's usually sweet and gentle and quite loving. He's not uncommunicative. But he does blow up occasionally and if he does he can knock you over.

How is his vocabulary today?

Schawlow: Pretty good. When my wife was alive he was very open with her, and would sit in the back of the car and type away and tell her a lot of things, some of them fantasies like jobs he wished he had but didn't really have.

Am I right in assuming he had been able to read and write for seventeen years, since he was 10, and you didn't know it?

Schawlow: Yes. He told us by typing that he'd been taught to read when he was 10 by Grace Turner. She was awfully good and ran a group home where he lived at that time, Cloverdale, 150 miles away from where we lived. When my wife asked why he hadn't told us before, he wrote on the machine that it was too hard. When she asked, "How is it you can do it now?" he gave her a sweet smile and typed, "I love you." Now he can't write without assistance. He still wants a hand on his wrist. We only see him every couple of weeks for an hour or so—it's 200 miles each way.

How many autistic people are there in America?

Schawlow: One in every 10,000.

There's no place nearer for Artie to stay?

Schawlow: No. It's discouraging. In 1990 an article was published in the *Harvard Education Review* about facilitated communication. The writer had seen a lady in Australia who was doing it with a number of autistic children. He came back and with great enthusiasm went around the country holding workshops to teach people. So facilitated communication has become a big thing. That's essentially what we were doing all the time. Several others discovered it too, independently.

Do you have a theory about autism?

Schawlow: Not really. I have a lot of theories about how people behave with autism. I think they're hypersensitive. They have great fear, and that's why they freeze up. And that's why the touch is so important, to make communication somewhat reassuring. I think there is some brain damage, but I don't know enough about the brain to know what it is. There have been serious studies by neurologists— even some brains have been dissected. They found the same anomalies, but no clear pattern as to just what it is. Some parts of the brain haven't been developed. I'm mainly interested in things I might do to help Artie. There the educationers and behavioral psychologists have been helpful. I think we're a generation away from medical help.

Does Artie realize your achievements as a scientist?

Schawlow: I think so. He knows a lot. I don't know where he picks it all up.

Does he continue to read?

Schawlow: Not a lot. He has one young man with him for 8 hours a day, and he reads with him. When he first started to communicate my wife would open a book and peek at it, then ask him a question. And he'd gotten what was on the page.

An autistic woman said, "If I could be non-autistic at the snap of the fingers I wouldn't do it."

Schawlow: That's strange. Most autistic people tell us by facilitated communication they wish they could overcome it. Very few have had successful careers. Some have shown remarkable ability. I met a man in the Orlando area in Florida who had a Ph.D. in math. He got a job teaching for which he was totally unfit. Since then he hasn't done anything much.

Are most institutionalized?

Schawlow: Ninety percent of them. And it's horrible. The one reason they don't progress is because nobody tries to teach them. Things are slowly improving. People are beginning to realize the institutions are terrible and are always going to be terrible. Some of the private places are terrible, too. When Reagan was governor he kicked a lot of people out of the state hospitals but didn't make any provision for them. A lot are living in the slums and subject to drug pushers and that sort of thing.

You're Charles Townes's brother-in-law?

Schawlow: Yes. As postdoctoral fellow at Columbia I met his younger sister, Aurelia, who came to New York to study singing. We were married for almost 40 years but then she was killed in an automobile accident on May 6, 1991. She was on the way to visit Artie and got about 150 miles and somehow plowed into a parked truck and died almost instantly. I've looked at the scene several times and you can go 100 miles along there and there'd be nothing parked. But this was in front of a convenience store and the shoulder lane looks broad enough, and I think she thought it was a driving lane.

May I ask, why was she going alone?

Schawlow: Well, it was a Monday and she just couldn't wait until I could go up on a weekend.

RELIGION

Is the Bible your favorite reading?

Schawlow: I don't read it very much, but if you ask me what I thought was the greatest book ever written I guess I'd have to say that was.

Are you religious?

Schawlow: Yes. I was brought up a Protestant Christian and I've been in a number of denominations. Lately it turns out that Artie wanted to go to church. When I go up there I've been going to church with him, at a very good Methodist church.

How does he respond?

Schawlow: Well, he just sits there. But he doesn't run away.

Did he tell you on his communicator that he wanted to go?

Schawlow: Actually we brought in a number of experts, and particularly Martha Leary from Toronto, who is very good at facilitated communications and a very warm personality, which I don't particularly have. He told her that he'd like to go to church. I didn't do anything about it for some months, but when we tried it he did like to go.

Do you believe Jesus was God?

Schawlow: I wouldn't say I disbelieve it. You know I'm getting old (72) and facing the prospect of death and you wonder: Could there really be some life after death? It's hard to see. There's no place in this universe for it. But is there something in some other universe? I don't know. I think it's better to believe. Certainly I think Jesus was the greatest moral philosopher. And the imitation of Jesus is the way to live your life, I think. Beyond that, I don't know.

You go further than most physicists. I find few who aren't out-and-out atheists.

Schawlow: A number are religious including Charlie Townes. You can't ever get proof or disproof of religion. You have to have faith and it's based on faith. Curiously enough, now I have a computer program that will search the Bible and the word faith only appears a couple of times in the Old Testament but quite often in the New.

But isn't the scientific creed that reason is superior to faith?

Schawlow: I think that faith shouldn't go against the facts. Reason is fallible just as much as faith is. One can make mistakes in reasoning. One other point. I think the idea of a personal God is very reasonable. After all, people who work with a time-sharing computer find it acts as though the whole machine is giving its whole attention to them. If a dumb machine, a time-sharing computer, can do that, this wonderful entity we call God can do the same. I find prayer to be helpful sometimes. I feel better. But whether it does anything . . . I haven't had a religious conversion experience, you know.

MASERS AND LASERS

How did you and Townes come up with the laser?

Schawlow: Townes has the idea for the maser (microwave amplification by stimulated emission of radiation) in 1951. In 1955 the first masers operated and the next couple of years there was a lot of progress. But the original idea had been to try to get the wavelengths shorter, in order to produce with electron tubes. And they had not succeeded. In 1957 I began to think seriously how one might go further. At that time Townes was consulting with Bell Labs and we had lunch together. He asked me whether I was interested in the problem. And we decided to study it together. At his suggestion we decided to jump over the infrared region, because not much was known there, and to look at the visible and near-visible region. I looked up the transition probabilities and lifetimes and so on, and found particular sets of atoms where I calculated we could get enough energy to pump what we called an optical maser.

The problem is that going to light wavelengths the atoms don't stay excited. They radiate very quickly. But you could get enough to get laser action. Then we had to invent a new structure, because the box he used just the size of a wavelength for the maser would be impossibly small for the visible region. If you could make one that small you wouldn't have room to put any atoms in it. So then I had the idea of just throwing away most of the box and using two little mirrors facing each other at the end of a long pencil-like column of material, and that would pick out a particular direction. So only the waves going back and forth between the mirrors would stay long enough to get built up much. Once we'd done that we realized we had the basic principle and published an article on it in 1958.

Have you done any work on it since then?

Schawlow: Probably the most important paper I wrote in the 1970s was one I did with Ted Hench in which I showed how you could use lasers to cool atoms to very low temperatures.

And the advantage of that?

Schawlow: You could get the atoms down to temperatures a fraction of a degree, make them almost stand still, so you could study them for a longer time. Now laser cooling is a very active field. They can trap the atom, and they make very sensitive devices for measuring gravity, for instance, more accurate than other gravity meters. They can measure the rate of fall of a free atom very precisely and better frequency standards and so on.

Is light still a mystery?

Schawlow: Some things are. We only know light from its interactions.

And electricity, too, is still a mystery?

Schawlow: Yes. I guess I feel I'll never really understand physics. I understand a lot of little bits and pieces. The whole picture is infinitely complicated.

To the scientist the laser is produced by a light being amplified by stimulating the emission of radiation. Can you explain this to the layman?

Schawlow: An atom in its normal condition can absorb light of certain wavelengths. So the light wave that comes in is lost, but the atom has for a moment some stored energy. It may hold that only for a millionth of a second, then it will emit it by randomly emitting a flash of light—what people call a photon. During the brief instance that it is excited, if another wave of the same frequency comes along it can stimulate or force it to emit. And it emits in step with the incoming wave. So the incoming wave is amplified. However, the absorption and the stimulated emission are the true negative of each other. That is to say stimulated emission is negative absorption. So if you have more atoms in the lower state, then absorption always predominates. At any finite temperature you always have more atoms in every lower state. So you have to get away from thermal equilibrium. If you can do that and get more atoms in the excited state then the amplification will predominate. And that is what was discovered in the last half century.

Is it difficult to do?

Schawlow: No. It turns out to be easy. We just hadn't realized the possibility. People were so imbued with the concept of equilibrium that you could never get far from equilibrium. They didn't think of the possibility. And it came about first through the radio frequency, or microwave, where the quanta are so small and so numerous that you couldn't help getting away from equilibrium. That's why I think the maser had to come before the laser essentially. Though it's easier to build lasers than masers.

Does it occur naturally in nature?

Schawlow: Maser action has been observed, but not laser action.

What is laser fusion?

Schawlow: That's usually having very intense beams of light, from a number of laser beams, pointing on this poor little pellet of heavy hydrogen. They fire this short pulse that heats and compresses it to a density and temperature greater than the center of the sun. So, briefly, the hydrogen atoms knock at each other so hard that some fuse into helium atoms and release energy. And if you can control that—like the explosions in an internal combustion engine—you can get almost unlimited energy. But so far the experiments require enormous lasers, and the threshold is a very high energy. So they're going to have to contain a very big burst of energy, when they do get it. But it doesn't look impossible. Even if it takes a century it would be worthwhile— providing energy for the world—electrical energy, lighting, heating, and comunications.

What's the most valuable use of lasers today?

Schawlow: Very, very diverse. The biggest use is for compact disc players. A great advance. Wonderful that we can hear music so clearly. There are huge numbers used for supermarket scanners. The most gratifying fields of use for me are laser medicine and surgery. I hope I never have one used on me, but they can do a lot of things that couldn't be done before.

I thought it was painless.

Schawlow: Not always. I thought it was, too. My wife had a doctor notice the beginning of some lesions in her eye which might lead to detachment, and tried to treat them with a laser. But she found it very painful. I'm told some people do, others don't. They use green light, to which the eye is very sensitive. Originally they used red light, but it wasn't very well absorbed by the red tissues of the eye. That wasn't painful. Mostly we use green today. The laser does things you can't do any other way. It will seal up the lesions and prevent detachment, because if the lesions in the retina get large the fluid will seep in behind the retina and lift it off the support, then you can't focus the eye. That's what's called detached retina. It's bloodless, that's the main advantage of the laser, and you can do it with a smaller opening.

How about on cancer?

Schawlow: If it's on the surface, they inject a dye which absorbs the laser light. It goes selectively to the cancerous cells and then they radiate with red light from a laser, and those cancerous cells will die. It has been used on lung cancers and some skin cancers. But most of the uses of lasers so far in surgery have just been in cutting and welding. They're useful in gynecology for precancerous lesions of the cervix. Also, when bleeding tissue gets outside the uterus and causes a lot of trouble. Lasers can remove that. The fact that it cuts cleanly and bloodlessly is useful, but it is just a knife. People are only just beginning to explore this, the photochemical effects of a particular wavelength of light perhaps in combination with dyes will be much bigger. But it's a large and difficult field.

What range does a laser beam have?

Schawlow: It spreads out like any other light, but if you use a big telescope run backward—then if the wavelength is a micron (one millionth of a meter) and the aperture is a meter—the spread would be only one part in a million. So you could go a million miles and the beam would be a mile across.

Is this why people associate it with the so-called death ray?

Schawlow: The idea of death rays came long before lasers. You can trace it back to Archimedes burning the sails of enemy ships with reflected sunlight. Ever since there were lasers, newspapers have shown drawings of big cannon-like devices shooting down missiles and so on. They're all fictional. Even now I don't think we have any lasers that are useful weapons. Although there have been suggestions lasers could be used to blind the enemy. They're useful for guiding weapons. But you know "Star Wars" was a big fake. I think they got sold a bill of goods, because a study by the American Physical Society said they wouldn't even know for ten years whether the powerful lasers they would need could be built, let alone put into Space and controlled. [See Chapter Nine conversation with Robert Jastrow for a different view.]

Were you and Townes asked for your advice on "Star Wars"?

Schawlow: I don't know exactly what Townes did. He does some secret work, which I don't. We were asked to serve on the review committee to review the report and they wanted it done under secrecy. I said I wouldn't do that, although I think I had clearance by then. But I didn't want them to put out a report saying: "Yes, this is going to work, but we can't tell you why." It didn't work out that way at all. Townes did review the report with all the secret information. I reviewed the published text, and decided it was all right.

You both concluded that "Star Wars" was not feasible?

Schawlow: I concluded that the committee was very responsible, and intelligent and had done a responsible, intelligent job, so I believed what they said—that it wasn't practical.

The Federal government also asked for your advice on how to protect the currency from counterfeiters. How did it go?

Schawlow: That's the time I got secret clearance. The report was secret but the main findings have been reported: that it's going to be an ongoing battle. What they're worried about are the color copiers. However, with the present generation of copiers they can't do really fine text. So I believe they decided to put some fine text around the borders of the higher denomination bills, around the picture. They also decided to put threads in the paper with something printed on them. They're not doing anything about $1 bills. One thing I invented which never became practical is the laser eraser, which would take ink off paper. I once had an FBI man ask me to erase a couple of samples, because they were working on a case where they thought somebody might have used a laser eraser. Then he asked me how criminals might use lasers. I said they might take the ink of one dollar bills and get special paper to print twenties. And he said, "Oh no, they wouldn't do that. They'd print hundreds." The hundred dollar bill is the most (frequently) counterfeited.

It could be used for altering wills.

Schawlow: Yes, but it isn't all that clear most of the time.

I know listening to jazz records is one of your hobbies. Do you watch much TV?

Schawlow: Maybe an occasional football game.

No news reports?

Schawlow: No. News on TV is too slow for me. I get it on the radio and I read four newspapers every day: *Wall Street Journal, New York Times, San Jose Mercury,* and the *San Francisco Examiner.* And I read maybe thirty scientific journals.

What are your politics?

Schawlow: I tend to be a moderate liberal and end up supporting Democrats, and occasionally give a little money to a candidate, either Democrat or Republican.

What do you think of Clinton?

Schawlow: He could be worse.

What was your father?

Schawlow: A life insurance agent in Toronto for Metropolitan Life Insurance. Quite a lowly job. He had to go around every week and collect a few cents from this one and a few cents from that one. A very hard job. He had to go out every evening because that's when people were home. My mother was a housewife. I have a sister who is married and still lives in Toronto.

Did your father encourage you to be a scientist?

Schawlow: Yes. He kept telling me I'd have to go to Germany to study. I couldn't have afforded to go to college if I hadn't won a scholarship. The same with my sister. My father came from Latvia and went to Darmstadt to study electrical engineering, but arrived too late for the start of the term. So he went to the United States where a couple of his brothers had already emigrated. He lived for a while in New York and married my mother, who was from Canada. And she persuaded him to move there. I don't think he would have been a good engineer, because he was no good at fixing things. But he had a considerable math ability.

A good father?

Schawlow: I think so. I could never beat him at chess.

Where do you live?

Schawlow: I'm on the Stanford campus, about $1^1/2$ miles from my work. The faculty ghetto they call it. You lease the lot and build your own house, or buy it from another faculty member. That's what we did.

Do you have many friends?

Schawlow: I'm a pretty unsociable person. I keep telling myself I should make more friends but I'm so busy all the time and at the end of the day I'm dead tired. I've had temporal arthritis for four years and am taking a lot of drugs that suppress my immune system. I just get awfully tired, and if I overtire I get sick.

If you were starting out today would you go for a different field?

Schawlow: I think the most exciting thing for the next generation might be the neurosciences. I don't know that I'd go into it. I'm a simple-minded person. But I think that it has great possibilities. Articles appear now, like in the last *Scientific American*, reflecting the views of those scientists who have worked on particle physics, saying physics is dead because they can't get to high enough energies. But I think that's only one frontier of physics. I think the frontier of complexity, understanding molecules, and light, and solid-state devices, is a rich frontier of physics.

Do you believe in the Big-Bang theory?

Schawlow: Tentatively. Astronomers are very brave and bold, and make vast assumptions based on very little data. Very clever, though. There seems to be a great deal of evidence for the Big Bang.

In a recent book Francis Crick's view of the mind is that everything is physical, that thinking and consciousness are simply millions of neurons firing in the brain.

Schawlow: I'd take the word simply out of there. I think the complexity you can get when you have millions of neurons could amount to almost anything. I have a certain humility about science. We know a lot of things, but we don't know everything. And there may be some surprises. I remember at the time of the 1981 Nobel prizes, the Swedish TV arranged a panel discussion among the science winners. And Roger Sperry, who is in brain research, argued that he didn't think quantum mechanics was enough to explain the workings of the brain. You need something additional. And physicists and chemists argued with him, but were really just expressing opinions based on the idea that this was a field of energies where we knew the laws pretty well. But with the enormous complexities there may well be organizational principles that we haven't dreamed of.

I'm surprised you weren't more adamant in response to Crick's theory, as you're religious and open to the possibility of life after death.

Schawlow: Well, look. I view it the same way as I view evolution. I don't believe in the simple-minded theory that it all started in seven days, 4,000 years ago. On the other hand, I don't presume to say how God would have chosen to conduct evolution. Whether there was some overguiding along the way, I don't know. I was just reading an article in which they point out some insects' eyes are very exactly corrected with forethought on the curvature of the lenses. That just happened. Our eyes refraction is limited, apparently. So it's so wonderful. I just don't know. And if I believed individual prayer existed I would be believing there's some form of communication which we don't understand. Certainly telepathy has been debunked many times, but I don't think it's totally impossible. I tend, I guess, as a scientist, to believe things until I can't believe them any more. I know a lot of people who just pooh-pooh every new idea. Rather, I say, "Could that be?" and look at it in various ways, and see if it all hangs together with everything else I know. Some ideas you reject at that point. But I tend to start with a positive point of view. And a good bit of humility. We don't know it all.

Who was the most remarkable scientist you met?

Schawlow: Charles Townes is pretty remarkable. Or Bohr's son. I never met the father though I heard him speak once. But I couldn't hear him. He had a very feeble voice and was very difficult to understand. I was impressed by his son's enormously broad knowledge of physics. He was at Columbia when I was there. He was running a theoretical seminar, and Von Neumann came and gave a talk on the theory of turbulence. And I don't think anybody understood it except Bohr's son, and he asked some penetrating questions. He seemed to know everything. Certainly his achievements, though great enough to win a Nobel prize, were not as revolutionary as his father's.

What's your most absorbing interest as a scientist these days?

Schawlow: Not much because I'm retired. I don't have any students. I'm fascinated by the nonclassical nature of light and the nonlocal nature of quantum mechanics. I've always thought of light as a simple electromagnetic wave, but it is quantasized as well. And you can set up various nonclassical states where the fluctuations are all in the phase and not in the amplitude, or the other way around. But the thing that really fascinates me is the experiments which show that things quite far apart can somehow know what's going on with the other. I think it really shakes up our ideas about space and time.

Weisskopf says the great distance apart isn't so, that the quantum sort of spreads.

Schawlow: Yes, that's okay. Except that in the experiment suggested originally by Bohm, in which they rapidly switched the detectors' polarization, so that the detectors are 14 feet apart—that's 14 nanoseconds, one foot per nanosecond—they switch the polarizers in 10 nanoseconds and yet still get the same correlation. So that some information must be transmitted faster than light. At least, that's my interpretation of it. Nobody knows any way to transmit information that way, because it's only the correlation between the two distant points that seems to go that way. With the individual ones you get quite random counts. But when you put the two sets of data together, the two detectors show the correlation. Fascinating. There are a lot of experiments going on. The experiment I'd like to see done is to increase the distance to maybe a mile or so. It's possible that quantum mechanics is only a somewhat local approximation, that whatever this is, propagates perhaps not at an infinite speed but at a high speed faster than light. It doesn't bother me that something goes faster than light. Because, after all, light goes faster than sound.

This would refute Einstein.

Schawlow: No, it would particularize Einstein into things involving electromagnetic interactions and not everything. Einstein himself was very clear about the peculiarities of quantum mechanics. He and Podolsky and Rosen in that famous paper I've never read, but is often cited, were the first to propose a thought experiment that may test whether quantum mechanics was right. Of course, quantum mechanics survived, but I'm sure Einstein would have realized that it's telling us something—that quantum mechanics is very strange.

Or as Richard Feynman said, "If people say they understand quantum mechanics, they're lying."

Schawlow: Yes. Dyson once said his experience was when people start studying quantum mechanics they say they don't understand it. After a couple of months they say they understand it, but they've just gotten used to it.

Now you're retired what most preoccupies you?

Schawlow: I put in a lot of my time into things for my son. Even this group home where he is went bankrupt and we had to take it over and set up a nonprofit organization and try to balance the books and get a new manager. And I've spent a lot of time defending facilitated communication. It's not yet understood how it works. With autistic people there seems to be a difficulty in starting or initiating motion, and touch seems to help that. Also they don't have good control over their muscles. Steadying the hand seems to help. There's a wonderful device from the Franklin Company, Language Master Special Edition. It has a little keyboard and also talks. If you type in a word and then press "Say," it will say it very clearly. Or you can preprogram up to 26 phrases you can then get by pressing a single letter.

Is Einstein your hero?

Schawlow: I wouldn't have used that word. But I think he's perhaps the greatest scientist that ever lived. A man I most admire in many ways is Faraday, who was a simple experimenter, but such a brilliant experimenter. These guys are just miles ahead of anybody else.

Chapter Twelve. Harold Urey

"I regret that some of my scientific experiments should have produced such a terrible weapon as the hydrogen bomb. Regret, with all my soul, but [I do] not [feel] guilt."
—Harold Urey

Harold Urey gets my vote for a Nobel prize in generosity.

During the Depression he got a grant of $7,600 from the Carnegie Foundation to reward him for his discovery of heavy hydrogen, also known as deuterium. "I looked for it," he said, "because I thought it should exist." His Columbia University colleague, I. I. Rabi, was struggling along with little financial support for his research. So Urey gave him half the grant money.

"What a greatness in Harold Urey—what a tremendous magnanimity to do something like that," said Rabi to writer Jeremy Bernstein. "It was one of the most extraordinary things imaginable. He had a deep faith in me and told somebody that I was going to win the Nobel prize. [Both men did.] The money set me free! It made me independent of the physics department," and free to pursue research fulltime.[1]

Urey won the 1934 Nobel prize for Chemistry, soon after his generous and unexpected gift to Rabi. Like the Carnegie grant it, too, was "for the discovery of heavy hydrogen."

Heavy hydrogen "was found to constitute one five-thousandth part of the earth's waters, including the water in the tissue of plants and animals. It was shown to have an atomic weight of two—double the weight of common hydrogen."[2]

Harold Urey started life as the son of a minister who also taught school. His father died when Urey was six, and his mother eventually married another clergyman. He first studied in rural schools in Indiana and then taught in them in Montana. Like many others destined for greatness, Urey also attended the Institute for Theoretical Physics at the University of Copenhagen, studying under Niels Bohr between 1923 and 1924.

While an associate professor of chemistry at Columbia he discovered heavy water, manufactured from heavy hydrogen, in which the molecules consist of one atom of oxygen and two of heavy hydrogen. Heavy hydrogen atoms were so rare that an amount of water equal to the weight of the entire universe would have been needed to produce one centimeter of heavy water.[3] Its discovery, ranked among the leading achievements in modern science, has greatly stimulated research in physics, chemistry, biology, and medicine.

In 1938 Urey reported the results of more experiments: "We now have [the heavy isotopes] of hydrogen, oxygen, nitrogen, and carbon—atoms from which 75 percent of all substances are formed." Two years later he announced his production of the isotope of sulphur—important for biological and medical research.

He kept canaries in the laboratory while working on the carbon atom to detect if any of the two poisons used were escaping [apparently none did]. And he spent many nights there, eager to monitor the progress of his experiments.

The first public scientific lecture Richard Feynman ever heard was given by Urey in Brooklyn when he discussed heavy water. They met again at Columbia University in 1942 when both were working on the Manhattan Project and Feynman was carrying "a tiny piece of uranium [to be weighed]. Wearing his battered sheepskin coat, he had trouble finding anyone in the building who would take him seriously. He wandered around with his radioactive fragment until finally he saw a physicist he knew, Harold Urey, who took him in hand."[4]

During World War II, Urey directed the Manhattan Project's separation of uranium isotopes. At first Urey thought his discovery of heavy hydrogen [deuterium] might have the practical value of, say, "neon in neon signs." Instead, it helped to create the hydrogen bomb, and heavy water proved a useful coolant in producing atomic bombs.

In 1943 Urey feared that the Nazis already had the atomic bomb, and any moment might drop it or large clouds of radioactive material on American cities. His warnings were taken seriously by General George Marshall who secretly sent Geiger counters to Washington, New York, Boston, Chicago, and San Francisco to detect radioactivity.

Horrified by the mass deaths and massive destruction caused by the atomic bombing of Japan, Urey joined those urging a reconsideration of the ethical problems involved in any possible future use of the bomb, especially in irresponsible hands. He was also for its international control. "Above all I regret that some of my scientific experiments should have produced such a terrible weapon as the hydrogen bomb," he said. "Regret, with all my soul, but [I do] not [feel] guilt."

Urey also pioneered the application of isotope research for use in geology and paleontology. His oxygen thermometer has enabled scientists to calculate temperatures on the earth millions of years ago, through studying fossils.

Perhaps his most intriguing experiment was an attempt to create life from nonliving matter by simulating primordial conditions on earth. He and Stanley Miller of the University of California at San Diego, supposed that life on earth began when lightning struck through clouds of methane and ammonia to create complex organic chemicals. They simulated this in a 1951 experiment by using electricity as the energy source, shooting electric sparks through a soup of methane, ammonia and water, and producing four amino acids— probable precursors to life. [More recently, chemical evolutionists, such as the University of Maryland's Cyril Ponnamperuma, believe life on earth may have sprung from a hydrogen-poor atmosphere similar to our own but without oxygen.]

Urey looked with fascination, too, above the earth, at the solar system, and in 1952 Yale University published the result, *The Planets: Their Origin and Development*.

He was an enthusiastic proponent of space exploration as a way to discover clues to the origin of the solar system, and to search for life on other planets. But his eager request to join astronauts on their trip to the moon was rejected, he was told, because he was too old.

When I spoke with him shortly before his death at 86, in 1981, he was professor-at-large of chemistry at the University of California in La Jolla.

As the son of a clergyman and stepson of another, were you religious in your youth?

Urey: That's right. But with my studies I became less convinced. I'd now call myself an atheist.

Do you believe the universe was created by accident?

Urey: It's a matter of natural law.

And we living creatures were created spontaneously?

Urey: I don't know where we came from. But to assume there's a God that knows all about it, is a big assumption.

How about some intelligence greater than ours?

Urey: I don't know why I should assume that.

In other words it's a mystery.

Urey: That's right.

Do you visualize your ideas?

Urey: I talk of my ideas to my colleagues.

To get ideas bouncing back?

Urey: That's right.

Do you get any ideas from your dreams or after sleeping on a problem?

Urey: No, I never did. I guess that has occurred with others.

When you were Niels Bohr's student, did he discuss his ideas with you?

Urey: He was a bad talker. He didn't speak good English and I don't think he spoke good Danish, either. He was a sort of inhibited talker. I listened to what he said, when he was lecturing and discussing things informally, but I didn't understand it very well. When he tried to speak to Churchill and Truman he had difficulty convincing them about his ideas in regard to the atomic bomb. [He wanted to internationalize it.]

As a scientist are you trying to discover the secrets of life?

Urey: No, I'm not working hard on the origins of life. I have a student, Stanley Miller, who does all the work on that subject. I let him do it.

I don't mean the secrets of biological life. I mean the secrets of the universe.

Urey: It's a challenge but I'm working on it. I think it's a bigger problem than I can solve.

What has driven you?

Urey: Fascination with specific things I was doing, all the enormously interesting puzzles.

Not the general mystery?

Urey: I'm interested in them both. But I was fascinated by studying specific problems and getting a partial solution to them. Probably the most interesting thing was finding how to get the temperatures at which rocks were laid down (during the earth's creation). Now I'm paying attention to what they're doing on Mars. I'm also interested in the energy problem. And I'm terribly afraid of nuclear power— the enormous expense of it, and the danger of something going wrong. I think it's been a mistake to go so heavily into nuclear power, and I don't believe it's cheap. We should use sun power. Also enormous saving could be accomplished if we tried to economize. We waste an enormous amount of energy. Ford had a thing where you turn off three cylinders when you reach 45 miles an hour.

Why don't you scientists concentrate on things like that?

Urey: Scientists are fascinated by what they're doing and they're not much fascinated by working on the engine of a car.

How do you relax?

Urey: My little hobby is to go to the beach practically every day and walk on the beach.

And think of how to harness the power of the waves?

Urey: No. I usually look at the little animals living in the waves and things of that sort.

It puzzles me why no scientist has harnessed the waves.

Urey: It's an engineering problem. The principles are all understood. If they used the money they spent to produce the atomic bomb, they could harness the energy of the waves.

Did you agree with the decision to take away Oppenheimer's security clearance?

Urey: No, I did not. Oppenheimer was a completely honest man.

Was he a victim of the climate of the time?

Urey: Otis L. Groves, I think. [General Leslie Groves was in charge of building the atomic bomb at Los Alamos.]

Didn't Oppenheimer's past Communist contacts frighten his critics?

Urey: I don't think so. I think it was just a crazy attitude that there was a bunch of Communists about. As a matter of fact, General Groves was very suspicious of me, and I had no sympathy for the Communists at all at any time of my life.

What did you study in the solar system?

Urey: Physical and chemical problems.

Do you have any opinion about the possibility of intelligent life in outer space?

Urey: I suppose there might be.

You leave it as an open question?

Urey: Yes.

Why?

Urey: There are many stars in the universe that should have planets similar to ours. If we found a planet with the same or very similar conditions to earth, then it is probable that life would evolve there.

Do you consider parapsychology a reasonable scientific pursuit?

Urey: Yes, I would say so.

Dr. Edward Teller told me he was a complete disbeliever in parapsychology and thinks it's a waste of time to investigate the extrasensory world. You don't agree?

Urey: I wouldn't think so, no. The point is that one can't follow all the branches of science effectively, so I just am limited in what I do. And I don't pay attention to parapsychology.

How do you feel about life after death?

Urey: I don't believe there's life after death.

As you don't conceive of an intelligent or purposeful creator of our universe, do you assume it happened by accident?

Urey: No, not at all.

How, then?

Urey: I follow the astronomers and their hypothesis of a Big Bang 50 billion years ago.

Chapter Thirteen. Hans Selye

Hans Selye was ridiculed when he suggested that patients diagnosed as dying from a variety of causes—heart disease, ulcers, even cancer—were, in fact, dying from a common cause. His revolutionary concept of mental and physical illnesses was "more viciously attacked than anything since Freud," Selye recalled. Then, when it took hold, he was compared with Pasteur, Ehrlich, and Freud as a genius in his field. Even Einstein sent his congratulations, comparing Selye's unified theory of medicine with his own efforts toward a unified field theory of the physical world. But it was a long battle before he was accepted and eulogized, and regarded as the world's leading expert on stress as well as a man who had developed a philosophy of living to enhance and extend life.

When Selye was a medical student in Prague, his professor asked him and other students to diagnose the illnesses of five patients. Selye noticed that regardless of their symptoms indicating different specific illnesses they all had a loss of weight, energy, and appetite in common. He concluded this was due to the body's physical response to such stressful things as fright, noise, and exhaustion.

After studying in Europe he got a Rockefeller research fellowship at Johns Hopkins University in Baltimore, and later at McGill University in Montreal, Canada where, he said, "I found that injections of the ovarian hormone stimulated the outer tissues of the adrenal glands of rats, caused deterioration of the thymus glands, and produced ulcers. The rats died. Later, I found that any artificial hormone compounds, and stresses, and any kind of damage, did the same thing."

His work confirmed his suspicion that illnesses in humans, such as asthma, heart and kidney disease, high blood pressure, peptic ulcer, migraines, ulcerative colitis, rheumatoid arthritis, could be traced to stress that led to an overproduction of adrenal hormones. "The apparent cause of illnesses is often an infection, an intoxication, nervous exhaustion, or merely old age," he pointed out. "But actually a breakdown of the hormonal mechanism seems to be the most common ultimate cause of death in man."

He concluded that aging was not so much a matter of how long you had lived, but *how* you had lived: that it was a result of the total wear and tear to which your body was exposed, "as though, at birth, each person inherited a certain amount of physical energy. He can draw upon this capital thriftily for a long, monotonous, uneventful existence or spend it lavishly in the course of a stressful, intense, but perhaps more colorful and exciting life."

Selye never stopped making discoveries, about which he lectured in ten languages. When asked what language he dreamed in, he began to take notice of his dreams and concluded that he dreamed in the last language he spoke before going to sleep.

When I spent an afternoon with this amiable and enthusiastic man, he had been up since 4 a.m., as usual. He began his 14-hour day as director of Montreal's Institute of Experimental Medicine and Surgery by riding a bicycle up and down nearby Mount Royal. He was in his late 60s and had an artificial plastic hip bone. He kept his original hip bone in a jar in his study, where we talked.

Hans Selye: I'd been educated by the Benedictines under very rigid conditions of Catholicism, and I found it extremely difficult to accept that I should not be an egoist, because I was and still am one. On the other hand, I didn't like being an egoist. So I had to find a solution. It wasn't until half a century later that I found this peculiar idea—and that is the altruism which you should practice according to most do-gooders, and which I call "altruistic-egoism." You see, you can't blame a big fish for eating a little fish, because if he doesn't he dies. And you can't blame me for being an egoist, because it's a natural law. Those who deny they're egoists are not introspective enough, or they're fooling themselves. You have to be an egoist but you don't have to be a fatalist. If you are always an *absolute* egoist you'll make many enemies. Then you are always in danger of being attacked.

Mightn't that be stimulating to some people?

Selye: A mild degree of it is. On a large scale, any disadvantage can be turned into a stimulus, because it wakes you up. In medical terms and on a larger scale, shock therapy, insulin shock, electroshock has been very effective in psychosis. It's extremely unpleasant, but it still wakes you up. Until we find better drugs we have to admit that unpleasant distress can be curative. Coming back to the idea of altruistic-egoism, I think you can marry the two in such a way that you can admit shamelessly that you are an egoist for your own good. My technique has been as a physician to heal. But I admit I am doing it for egotistical reasons.

How do you justify scientific research that could be used for harm?

Selye: Any new knowledge can be used for good or bad. I refuse to inhibit scientific progress on the basis that somebody could use it for something bad. I agree with those who oppose the use of atomic weapons, but the human brain is curious and wants to know more about itself and its world, and I'm all in favor of such research despite the fact that it can be ill used. My altruistic-egoism is an answer to your question. Most codes are based on commands, what you should or should not do. Nature does not proscribe. But the laws of nature are solid. That is to say, water boils at 100 degrees, and there's nothing you can do about it. I have developed a lot of natural laws concerning stress. These laws are forever. I can't imagine, for example, anything which will prevent cortisone from inhibiting inflammation. And I wondered whether one could not bring up to date the old, mostly religious but partly philosophical beliefs that you should do this, and shouldn't do that.

It's important to understand the principle that you can still be an egoist and a hoarder, which is very typical of egoists—they want to hoard money or power or position. But you can be an egoist without hoarding any of these. I am a hoarder, and I am a capitalist by conviction. But I have decided that the most permanent and valuable currency that is not subject to devaluation is love. Not my love for you, nor your love for me. I am using love in the biblical sense, which includes good will, and being useful to people. Instead of accumulating a fortune in money, you accumulate a fortune in your social contacts and standing.

One objection to my altruism-egoism gives me the greatest pleasure, because everybody says: This is what we Jews, or we Arabs, taught all the time. Recently I had a rabbi from Seattle who came to the conclusion that my idea is what Judaism has always taught. The Dalai Lama's secretary wrote to me saying it's perfectly in accordance with Buddhism. He reproached me though, for the fact that it is so unoriginal. And that's what I like about it. Because the others are creeds, and I'm trying to show that these creeds have a biological basis. I am usually blasted by Catholics who are so convinced their belief is absolutely true that they say, "I've got Jesus, so I don't need you."

First they say they don't need me and second they say you have to do exactly what Jesus said: "Love thy neighbor as thyself." That's a command. I say you can die on command on the battlefield, but you cannot love on command. Now "love thy neighbor" is generally considered a Christian principle, which the rabbi explained to me is also in the Old Testament. The Maharishi Yogi, another friend of mine—you wouldn't think a professor of experimental medicine and surgery has friends in those strata [He chuckled]. But I love the guy. He sees things from an entirely different point of view, but we come to the same conclusions. I'm not contradicting this law of "love thy neighbor." I'm just saying you can't do it on command. I don't want to change the meaning, but to put it in the language of today which is the language of science. Even the vicar accepts that physical laws work.

So, for me a thing is worth doing if it earns my neighbor's love. Throwing somebody out who's bothering everyone at a party won't earn the thrown-out one's love. But then you have to distinguish what is worth doing. Einstein said that if everybody says no and one person says yes, he's in a majority of one. It isn't necessarily how many people object to him, but what kind of people, and what service he has rendered. You cannot by every action please everybody but the lietmotif is nevertheless to earn your neighbors' love, the majority of ten neighbors' love, the most important neighbors' love, and so forth.

According to Auguste Comte, all knowledge goes through various stages and the first is always religious. Even in prehistoric times people believed in divine spirits—because something had to be created that was superior. To me, nature created man; and nature is superior. I don't require an anthropomorphic God with a long beard and looking very powerful. So I go back to what I believe is the source of all human conduct, which is to be natural.

What about the negative, destructive side of nature, the law of the jungle?

Selye: We are superior to animals in that we have intellect, and if we use it and have self-control, it can help overcome certain instincts. Owing to evolution man has developed such perfection of the central nervous system that we can have purely reflex responses, or purely emotional responses—including cruelty—subjected to the control of logical analysis.

I think it's impossible to be an absolute altruist because it's against nature. But you can be an altruist in the sense that you always consider, How much harm do I do to others? I remember when my library burned, because this made a great impression on me. It gave me great satisfaction that I got in the same mail a $75,000 check from the United States Government and $1 in cash accompanied by a letter on lined school paper written in pencil by a French-Canadian up north in Quebec: "I know your library burned and one dollar isn't going to do you very much good. And we haven't much money, but whether we eat meat tonight or just vegetables doesn't matter, and it's just one dollar, so let me contribute to something." I thought that was so touching. I almost cried.

Now I don't think it's biologically possible to love my neighbor as I love myself because somebody commands it. On the other hand, if by the philosophy by which I have lived very happily for 66 years, trying to act in life in such a manner that I *earn* my neighbor's love, and that in the broadest sense, earn their gratitude and respect by being useful to them, that gives you tremendous power. Much greater than political power, because your party may go under. Much greater than monetary help, because the dollar may crash. And I think that somebody who is useful to everybody, like Einstein or Sir Alexander Fleming, is in the strongest position, far stronger than multimillionaire businessmen.

Do you think though, that Fleming [who discovered penicillin] was acting out your philosophy of earning his neighbor's love?

Selye: He earned it, whatever his motive. That isn't pertinent. For example, a fellow who is desperately trying to accumulate wealth because he wants to assure his security no matter what happens is going to be very disappointed if the dollar loses value. Whereas, if you have earned your neighbors' love by being an awfully good writer of a book, you will have so many friends and so many people will want you to prosper that your safety will be that much greater.

You are in fact using modern science to confirm religious and philosophic teachings.

Selye: Exactly. And I try to make it clear that this is not a deviation from philosophy but quite independent of it. That is to say, I am not examining, as most religions do, the existence or nature of the Creator, or the purpose of creation. I start much lower, as if a mechanic has got a machine and doesn't know who made it, and why. He just wants to know how it works, so he can repair it if it goes wrong. As a physician, my machine is the human body. And my whole philosophy is based on the laws of that machine. So it is quite compatible with, and can be a complement to, any religion. It's certainly not a contradiction. I believe that the earning of gratitude from your fellow men by actions of yours which make you desirable and which eliminate all motives of aggression is a powerful way of maintaining what we call technically "homeostasis," that is to say, your steady state, your safety in society, and in yourself. And I try to base this on hormone determinations and measurements.

[To illustrate that his philosophy is based on nature, Selye pointed out that even primitive life forms have a balance between altruism and egoism; and that to produce stronger more complex beings, individual cells gave up their independence.]

STRESS

How does someone cope with a man in power whose own comfortable stress is to cause excessive stress in others?

Selye: He quits. You cannot caress a porcupine. After I first said that I received at least ten pictures of people caressing porcupines. If you're nice to him a porcupine will put down his quills, and then you can caress it. Since that time I've thought of a better example: the crocodile. You cannot caress a crocodile. There are bosses who are crocodiles. In a case like that, the only thing to do is to fire yourself.

So, in *Death of a Salesman*, the father should have quit his job?

Selye: Yes. And it's what you do with a boss who is a psychopath, a homicidal maniac. I emphasize: The code that you live by should be the best code, but not the perfect code—because that's unattainable. Affection is attainable. The crocodile is the evolutionary equivalent of the sadist and homicidal maniac. I don't think the crocodile has the brain to cooperate. On the other side of the coin, a three-month-old puppy is always caressable because his brain is developed that way. Even if you get mad at him and slap him, he'll go away and whine a bit and later he'll come back and snuggle up to you whether you were just or unjust to him. A three-month-old puppy to the crocodile is a range of psychological activity which you can find every day right here in Montreal, in the human colony. I try to be nice to everybody above the crocodile level. [He chuckled.]

Have you made any recent discoveries about stress?

Selye: Many. For example, hormones which increase resistance. I wrote a two-volume monograph on it.

In your book, *The Stress of Life*, you quote a Professor Peter Forsham commenting on the dissociation of the ego and the id. He tells of a patient under hormone treatment who could play the piano brilliantly until she was taken off hormone treatment when she reverted to her usual style, playing poorly. Would the effect of hypnosis be identical, in that people under hypnosis can be persuaded to perform more effectively?

Selye: It would be a bit daring to say that was similar or identical to hypnosis. It's an empirical fact.

But in both cases the hormone treatment and the hypnosis seemed to remove the inhibiting part of the mind.

Selye: I would have interpreted it the same way, but I have no scientific proof on that.

Is hypnosis a mystery to you?

Selye: The biochemistry and mechanism of it are certainly unknown.

If, as you say, physiological aging is determined by the total amount of wear and tear to which a body has been exposed, is it true for men and for mice alike?

Selye: Yes, but there's the added factor for man that if he's aware of this he can prolong his life, not by leading a less exciting or a less wearing life, but with a positive mental attitude. The fact that a man knows this enables him to change his fate by using what I call adaption energy, or adaptability. How you're going to roll with the punches.

Are the effects of childhood stress or traumas irreversible?

Selye: Whenever you're exposed to stress, a number of biological reactions go on in the body, and 99.9 percent of them are reversible. But there is only one minute fraction that leads to irreversible results: what I call chemical scars. If a thousand reactions go on and all lead to soluble ends products, homeostatic, self-regulatory responses will put you back to normal again. If you eat too much sugar, you will burn up the excess, or excrete it in your urine and be back where you were. But out of millions of reactions throughout your lifetime, one very minute proportion is irreversible, in the sense that it leads to an insoluble precipitate which cannot be taken away, so that's the scar.

Then you agree with Freudians who say childhood experiences are irreversible, but knowledge of them can help overcome future stress and help you to adapt to life.

Selye: Yes. You see, if you are aware of certain things, the awareness itself will often lead to subconscious defense-mechanisms, which you put into effect. In my own case it helped me an awful lot to live in such a way that a lot of people have reason to want me [He chuckled] to go on doing what I'm doing.

Do you recommend any chemicals to alleviate stress?

Selye: Tranquilizers are very useful when there is sufficient reason to give them. I'm much more in favor of developing the right psychology and mental attitude to stress, rather than give any drug.

Psychoanalysis, hypnosis?

Selye: No, not necessarily. You have to develop a way of life which teaches you to avoid stress, though I don't advocate trying to avoid all stress. Stress is part of life, but you have to know how to live with it. People who try to always avoid stress become vegetables. It's not worthwhile to prolong life by not using it. But you can use it intelligently, and avoid wasting vital energy.

You said: "Variety of experience is not only the spice of life, but possibly the key to longer life." You also point out that one can't completely avoid stress. But sustained stress can be fatal.

Selye: Sustained is not the right word here. Stress is sustained from birth to death. For as long as you live, there is stress. It's a matter of whether it always hits the same organ, whether it's sufficiently varied, whether you know how to react to it. The same news which may be terribly stressful and disagreeable to you, may be particularly relaxing and useful to me. If we have a bet on something and you lose, we get the same news but to me it means something else than to you. If it is very stressful to conduct an interview like this, then it is very bad for you. But if you get fun out of it, then it's relaxing.

How about those who say they like stressful situations, having their backs to the wall?

Selye: I like stressful situations. But you have to know how to use them. It's the same with gymnastics. That's physical stress. If a man who is not physically fit gets beaten around it doesn't do him any good. But if he's a professional athlete, he needs it to limber up. It depends largely on genetic factors, on your ancestors. If you come from a family and have inherited their genes which are good for a rapid pace of life, being forced to live a sedentary life is very unpleasant. But if it's the reverse, then the result is also bad.

I notice you keep your hip bone in a jar over here.

Selye: I've got a plastic hip. My hip caused me so much trouble and pain, it gives me vicarious pleasure to see it there every time I walk by. The friend who did the operation, an orthopedic surgeon at Cornell, has done 2,000 operations of this type and he says I'm the first patient who wanted to save the bone.

How do you define stress?

Selye: As the nonspecific response of the body to any demand. That includes both agreeable and disagreeable demands. There is the imaginary case of a woman sitting quietly in her living room reading a newspaper and she gets a telegram that her boy has been killed in Vietnam. She's under terrible stress as a result. These are the only indicators of stress which the public will notice, but not the biochemical changes: First she will get up. She wouldn't go on reading the newspaper. She will walk up and down, although there is no particular place she wants to go. Her heart will beat faster. I let her rest for a few months and she gets over it. And again she's reading the same newspaper under exactly comparable conditions and her son walks into the room in perfect health, because the news of his death was false. What will she do? Will she go on reading the newspaper? She will walk up and down. She will not only embrace him, she'll be overexcited generally, and she will have an increase in blood pressure and increased pulse rate. Her reactions will be almost exactly the same as when she thought he was dead. Both are stress situations. Stress is not necessarily a bad thing and the good kind we call "eustress", from the Greek preface "eu-phoria." So eustress is what she's experienced when the boy came in alive, and distress when she received the telegram. But as far as the body and mind are concerned, the two are experienced with equal intensity.

You may need a certain amount of electricity to cool a room by an air conditioner, and the same amount to warm it up—so that both are due to the consumption of electricity. But the results appear to be completely different. One person will react one way and another will react exactly the opposite way to the same event. That is why you have to be careful in interpreting as absolute values these questions that divorce is stressful and marriage is substressful, for example. It depends on the situation. A divorce will be experienced as an extremely painful event because you loved your wife. Or you will be glad to have gotten rid of her. So that to say divorce is stressful is meaningless.

How do you handle stressful situations?

Selye: I avoid them. That's part of my philosophy. First you have to have the motivation to know what you want to accomplish. And you analyze the event. Is that in my line? Is that worth fighting for? Can I do it, or not? If I can, am I willing to fight very much for it? If I can't, I give it up. At public speaking I'm pleased if I find a particularly receptive audience. I have eustress; good stress. If you are apathetic or half asleep you can't give a good talk. What bothers me, for example, is if I give an interview like this and you tell me at the end it didn't register.

[After checking the tape recorder] It's okay.

Selye: I've been working extremely hard since I was a student, but I love to do this. For me it isn't work, it's play.

This is from *The Forseeable Future* by Sir George Thompson: "Do you feel in the face of genius that it alone thinks naturally, while the rest of us block our thoughts perversely with irrelevancies." In other words, Einstein is the natural man.

Selye: I've worked on the psychology of research and of creativity. One outstanding characteristic of genius is that they look at things from new angles. Take an everyday example. You see the sun rising every morning and setting every evening, and you can see how it moves around. To think of it not moving around, that it is you who is moving around, takes genius. Most scientific discoveries are based on the ability to be free of prejudice and fixed ideas, and be prepared to think, "Now what evidence do I have."

So we can at least imitate genius by looking at the opposite of everything.

Selye: I do that. Whenever I have a theory about anything, I turn it around almost like a reflex, to see the reverse. For example, if you think of a disease as being due to X's production of something, it's useful to think that perhaps it's the shortage or using up of that same something that's responsible. You have to keep your mind open for unusual ways of looking at things.

Do you have any contact with foreign scientists?

Selye: I have a Russian working for me from the Soviet Department of Medicine, a Czech from the Czechoslovakian Academy, two Hungarians, a Swede, a Korean, and two South Americans. We're expecting a German in a few days and two Italians. At one time I had a Russian from the Soviet Academy, a dedicated Communist, in the same lab as a Hindu from India and a Catholic nun from Texas. They were quite tense for the first few days, then they became the most intimate friends and worked beautifully together. They even published together. So I think that's most encouraging. A few weeks ago sitting in your seat I had the head of the Soviet Authors Association who is part of the group that caused Solzhenitsyn all that trouble. He was very intelligent. I thought he would be hostile because I had signed a letter which was published in our Canadian journals supporting Solzhenitsyn. Because of all the information one gets about Communist thinking and philosophy, I was surprised to find him extremely cultured, and very well versed in philosophies, even physiology. He knew the biological aspects, and knew most of my books which had been published in Russia. I told him that the beaver is a typical capitalist, because he hoards materials for his own pleasure and my book could be used as an argument for capitalism. I'll never forget the look he gave me. By then we were on friendly terms, and he said, "Your kind of capitalism is the only kind that we Communists will ever accept."

Now for some slightly wild questions. In science fiction they often explore the possibility of cell regeneration, of human beings growing limbs that have been amputated. Is this even feasible?

Selye: I don't think there's a reasonable hope for it. It's possible in lower vertebrates, in amphibians. But not in mammals. Since it's possible in lower vertebrates there is a remote possibility we'll find out how the amphibians do it, and it might be applicable to man. That remote possibility exists, but it's better to keep one's predictions within [He laughed] reasonable limits.

Jean Rostand sympathizes with the plan to put dead people into deep freeze to be brought out when a cure for the disease that killed them is found. What do you think?

Selye: It's as close to science fiction as it can be.

But someone has already organized such a deep freeze.

Selye: Yes, he came to see me. He organized a society for this, but I think it's sheer optimism. I don't like to laugh at extreme optimists, I just like to appraise their chances. Even a chance in a million is worth a chance in a million, but I think you mislead the public if you mention these things too strongly, because they are not able to differentiate between a good chance and a very poor chance. You've got so little to go by to hope that this will be applicable to man, that I wouldn't bet on it.

Did he say if he got anyone to sign up?

Selye: I don't know. I received an awful lot of mail from people throughout the United States enquiring where this is happening. And I had to say I don't think it is anywhere.

What do you think about the monkey glands Somerset Maugham and I believe, Churchill, were supposed to have had injected for rejuvenation and to delay aging.

Selye: It was very fashionable after the first World War. The German Kaiser reputedly had this done when he was in exile. The difficulty is that then they didn't know enough about immune reaction to realize that these foreign tissues would be rejected, and not grow. A monkey gland does not grow in a human being. So the best one could hope for is that it might act as testosterone would. A monkey's testes contain testosterone, but it contains so little that nowadays this approach would be completely outdated. Today you can inject pure testosterone. In fact, we have anabolic steroids which are much more active than testosterone.

Would you inject yourself with any of these materials?

Selye: Not now. And I wouldn't recommend it for others before I was willing to inject it into myself. Take penicillin. When it was first discovered nobody would have wanted to inject it into human beings because the crude penicillin culture that Fleming had was highly toxic.

Are flying saucers outside your field of interest?

Selye: They fascinate me. I think they are definitely a possibility.

Have you heard the theory that they might be plasma in space?

Selye: No, but there's a book about flying saucers dedicated to me, by Jacques Vallee.* I don't know him. Never met him. In his book I am quoted three times as showing great understanding for this. And I must admit I don't understand it at all. Why did he have such an enormous success? I think it's double talk. When you talk to people who love him dearly they don't understand him themselves. It's almost like hypnosis.

How would you react if someone told you they'd seen a flying saucer?

Selye: As a scientist I always feel that I have got to give him an unbiased hearing. A thing quite contrary to everyday experience I'll never deny unless I have proof that it does not exist. All I can say is the chances may be very low but I wouldn't consider it impossible. A scientist has to be careful about what he says about his own field, because what he says is supposed to be based on evidence. It is not an opinion. The fact that someone is a scientist, say a pathologist, does not make him any more competent to judge flying saucers than anybody else. And one feels like a cheat, using the prestige one got by working in another field to add credence to what you say in a field in which you are no more competent than the person asking the question.

---

* Dr. Jacques Vallee was an associate of Dr. Allen Hynek, chairman of the Department of Astronomy at Northeastern University and scientific consultant to the U.S. Airforce on Project Blue Book—an investigation of unidentified flying objects.

What could our life expectancy be in the future?

Selye: We're guessing, but according to some verified reports there have been cases of people who lived to 150.

But everlasting life on earth is just too far out?

Selye: I think so. If you get to these farfetched ideas you can say almost anything. Do I think you could turn into an elephant? I won't say no, because there is no proof that you couldn't.

Will astronauts or space travelers die young because of stress?

Selye: Space flight is connected with very strong stress reactions, no doubt about it. And in space medicine, research on stress plays a very important role. I think that is one of the best established difficulties man will encounter in space. We will be faced with considerable stress.

Who makes most use of your theory?

Selye: They teach my code, altruistic-egoism, in schools and in nursing courses a great deal. The World Health Organization and UNESCO have published articles about it in their journals, approving it and trying to disseminate it.

Who is the author of the aphorism in your hall?

> "Neither the prestige of your subject and
> The power of your instruments
> Nor the extent of your learnedness and
> The precision of your planning
> Can substitute for
> The originality of your approach and
> The keenness of your observation."

Selye: Your obedient servant, myself.

You didn't sign it. Where's your egoism?

Selye: I felt that would not only be egoism but bad taste in my own institute. They also gave me a bronze plaque because of my 60th birthday. I left it in a box in the cellar, because I think that should be put up *after* my demise.

Chapter Fourteen.
Henri Ellenberger

"Take nothing for granted."
—Henri Ellenberger

Son of a Swiss Protestant minister, Henri Ellenberger was born in South Africa in 1905. He studied medicine and psychiatry at the University of Paris, where he got his M.D. In 1934, he practiced neuropsychiatry in Parisian hospitals. During World War II he was assistant director of a Swiss mental hospital in Schaffhausen, where he became friendly with several psychiatric pioneers including Ludwig Binswanger and Carl Jung. There he wrote a history of psychiatry in Switzerland, and papers on witchcraft and voodoo. From 1952 to 1959 he taught psychiatry at the Menninger School of Psychiatry in Topeka, Kansas, where he wrote a biography of the originator of the ink-blot test, Hermann Rorschach. For three years after that he taught and did research at McGill University in Montreal, and in 1962 became professor of criminology at the University of Montreal.

"Take nothing for granted," is Ellenberger's watchword. So, wanting to write a history of psychiatry and psychoanalysis he traveled to Austria, Germany, Switzerland, France, and England to question those with firsthand information. In Zurich he interviewed Jung, Jung's relatives, associates, and students. And he recorded their memories of Freud by the Reverend Oskar Pfister and Dr. Alphone Maeder. In London he spoke with Freud's son, Ernst, and Freud's colleague and official biographer, Ernest Jones. Ellenberger claimed to have found that 80 percent of Jones' "facts" about Freud were completely untrue or greatly exaggerated.

Tracing the study and manipulation of the mind from the exorcists, magnetists, and hypnotists to Janet, Freud, Adler, and Jung, Ellenberger published his "take nothing for granted" findings in *The Discovery of the Unconscious*, Basic Books, 1970. To learn more about Freud and Jung as well as Ellenberger, I spoke with him in his Notre Dame de Grace home.

You believe that both Freud and Jung suffered from what you call "a creative illness," as a direct result of which they made great discoveries about the mind. What form did Freud's creative illness take?

Ellenberger: He suffered from heart ailments, depression, and from psychosomatic ailments. The symptoms came and went. They were sometimes agonizing. He thought he was about to discover a great secret, perhaps by analyzing his dreams. But then would be tormented by doubts. He brooded constantly over past memories, and he felt utterly alone. He came through the ordeal exhilarated, sure that he had discovered a great truth, and instead of being unsure of himself he was confident. His personality had become permanently transformed. He was now convinced that infantile sexual experiences were the cause of neurosis. People say that after Freud recovered from his six-year creative illness in 1900, he spoke of the Oedipus complex and the libido with conviction as if they were absolute, universal truths. His enemies, however, said that Freud was simply neurotic, and that psychoanalysis was the expression of his neurosis. And his supporters that he had engaged in a heroic feat of self-analysis which revealed for the first time man's unconscious mind.

Did you coin the term "creative illness"?

Ellenberger: I think I did, but you never can be sure.

Is Ernest Jones' authorized biography of Freud fairly accurate?

Ellenberger: Many things taken at face value are not true. When Jones writes for instance, "Preiswerk said it was necessary to learn Hebrew as the language of heaven," I believe this was said as a family joke. Jones also wrote that Freud's enemies said that the only women who went to him were prostitutes. Again, I believe this was said as a joke. The joke was in the form of a question and answer: "Why do women go to Freud first and then to Jung? Because Freud has emachen [filles de joie], and the other are junfrau [virgins]."

You met Ernest Jones.

Ellenberger: I spent an afternoon with him in his home in a London suburb. He had a long table with papers in piles—Freud's documents. Some of them won't be released for publication until 200 years from now.

Is there much more to learn about Freud?

Ellenberger: If we can get the truth about Freud's childhood, I expect surprises. Frau Renee Gicklhorn told me that in Austrian archives there is an account of Freud's Uncle Joseph—the black sheep of the family—being implicated in a conspiracy concerning false Russian money. False rubles were printed in England and sent to commercial institutions in Europe. Uncle Joseph was arrested by the police while trying to sell three hundred or so false rubles to another man. He was sentenced to ten years in prison. Ernest Jones tells it as a slight matter: just a few weeks in prison. But it was much more than that. He was in prison three or four years and then he was pardoned because he told everything about the other counterfeiters in the conspiracy. Freud was ten years old when it happened, and it is not impossible that people talked about it. Frau Gicklhorn's hypothesis is that Freud suffered nervous shock, was teased by his comrades, got psychic trauma, and believed Jews were persecuted everywhere, although at the time there was very little anti-Semitism in Austria.

Freud was fascinated by the secret lives of others, yet reluctant to answer personal questions about his own life, and almost dreaded the thought that someone might write his biography. Can you explain that?

Ellenberger: That was the psychology of Jewry at the time in that country. And he, too, felt a great sense of privacy, despite the fact that he was penetrating the human mind. I would say it is rather a common phenomenon. You also have the old example of Virgil ordering that all his manuscripts be burned. Fortunately the emperor interfered and forbade it.

Who else suffered—and benefited—from creative illness?

Ellenberger: Franz Mesmer, who, when he recovered proclaimed that he had discovered animal magnetism. Nietzsche emerged from his creative illness with the philosophical ideas of "the eternal return," convinced that he was bringing a new message to mankind, a belief that life is eternal, that we re-enact every moment of our lives eternally. Nietzsche had a great influence on Freud, who said that Nietzsche's philosophical guesses and intuitions often agreed with his findings of psychoanalysis. Jung, too, was greatly influenced by Nietzsche.

What occurred during Jung's creative illness?

Ellenberger: It was not unlike Freud's self-analysis. But where Freud used free association, Jung encouraged unconscious imagery to reach his conscious by recording his dreams every morning, and using what he called "active imagination"—directing his daydreams and writing down what he had imagined. During his illness he believed that ghosts had invaded his house and told him they were the souls of the dead from Jerusalem. He had conversations with a feminine subpersonality whom he called his "anima."

Some people might have said he was going mad.

Ellenberger: No, I don't think it was psychosis. But he knew what he was going through was dangerous. He kept in strong contact with reality by fulfilling his duties to his family and his patients. His creative illness lasted from 1913 to 1919, when he emerged a new man with a new teaching. From that time on he spoke of "the anima," of "archetypes" and of "the collective unconscious" with absolute conviction of their existence. His life work and Freud's were a direct consequence of their creative illnesses.

Did anyone notice if Jung looked or behaved differently during his creative illness?

Ellenberger: Reverend Oskar Pfister, who was well acquainted with Jung, said he looked rather distracted.

In a BBC interview Jung told John Freeman: "There are these peculiar faculties of the psyche; that it isn't entirely confined to space and time, you can have dreams and visions of the future, you can see around corners. It's quite evident they do exist [these faculties], and have existed always. When the psyche is not under the obligation to live in time and space alone . . . then to that extent the psyche is not subject to those laws, and that means a practical continuation of life, a sort of physical existence beyond time and space."

Ellenberger: Yes, Jung was convinced there was another world that sometimes appears in this one.

What were his views about life after death?

Ellenberger: He thought it possible to communicate between the living and the dead, but that the dead needed us to answer their questions, not the other way around. He didn't visualize the next world as a place where blissful spirits were freed of all pain and suffering but he thought that, like our own earth, it might be a place where evolution occurred. He said, of course, that these were suppositions; that one cannot be certain.

You met Jung.

Ellenberger: Several times. I read first of all his writings and made a list of the points which seemed obscure to me. Then I interviewed him several times. He answered very patiently and extremely well. After that I wrote a summary of his ideas and teachings and sent it to him. He sent it back with extensive penciled annotations.

What was he like?

Ellenberger: A vigorous, muscular man. You might think his ancestors were peasants by the look of him, although that was not so. They were intellectuals. In contrast to his sometimes ponderous writing, his conversation was brilliant, subtle, fascinating, and profound. He had the gift, invaluable in a psychotherapist, of speaking with equal ease to people from all walks of life.

What were some of the things he said to you?

Ellenberger: He said "unawareness" is the greatest sin, and that it also is the cause of many neuroses. He believed everyone should possess his own house and garden, be active members of the community and, if religious, follow the commands of the religion. He told me that he went to Tanganyika in 1926 and lived there some months with an African tribe. He thought it would increase his knowledge of the unconscious mind by mixing with primitive men. He spoke with a medicine man from Senegal, a Christian maybe, who said to him, "I know there is a power of the devil because I have been hexed." Jung kept a diary of these conversations. He was specially interested in *The Tibetan Book of the Dead* [see conversation with George Wald in Chapter Seven] and wrote an introduction to the German edition. He thought the psychological knowledge of the unknown authors was remarkable, and compared the journey through the abode of the dead to what he had experienced during his own creative illness, but in reverse.

Did Jung believe in God?

Ellenberger: He said he didn't believe. He *knew*.

I understood that Jung had a personal reason for rejecting the Oedipus complex.

Ellenberger: Yes, because he was afraid of his mother and thought her homely. The idea of every son being in love with his mother seemed absurd to Jung. According to people in Zurich who knew Jung's mother, she was arrogant and not liked by the parish. But the parishioners loved his father. Freud's mother was beautiful, and he was his mother's beloved firstborn.

Theodor Reik believed Jung was prejudiced against the Jews, and quotes Freud as saying that he hoped Jung would give up anti-Semitism. According to Freud, Jung never did, welcomed Hitler, and became a dedicated Nazi. I have also heard that charge from others.

Ellenberger: I am absolutely sure Jung [a Protestant] was not an anti-Semite. But people were accusing him of it to the end of his life. As well as his father, his mother's father was a Protestant pastor. Sometimes I think Zionism is an invention of Protestant sects. They produced pamphlets saying the end of the world would be preceded by two events—first a Jewish state when Jews would return to their own country; second, they would convert to Christianity. This was supposed to happen soon. Jung's maternal grandfather, Samuel Preiswerk, was a distinguished theologian and professor of Hebrew. He was convinced that Palestine should be given back to the Jews. Jung's grandfather Preiswerk, was very enthusiastic about the first Zionist International Congress in Basle. Many people invited delegates from the Congress to their homes, and the Basle authorities gave them a room gratis. That is seldom done in Switzerland.*

---

* In a letter written to B. Cohen on March 26, 1934, Jung wrote that he was "absolutely not an opponent of the Jews, even though I am an opponent of Freud." On December 19, 1938 he wrote to Erich Neumann that everyone was "profoundly shaken by what is happening in Germany. I have very much to do with Jewish refugees and am continually occupied in bringing all my Jewish acquaintances to safety in England and America." *C.G. Jung Letters*, eds. Gerhard Adler, Aniela Jaffe, and R. F. C. Hull (New Jersey: Princeton, 1973), pp. 154 and 251.

Chapter Fifteen. *Theodor Reik*

*"Freud never indulged in the belief that he had solved all the riddles of the
inner life. He likened his work to that of the archeologist who had rescued a
few temples from the dark earth and brought them to light; but he had no
doubt that great treasures still remained below, awaiting excavation."*
—Theodor Reik.
"I have read your book [Reik's Listening With the Third Ear*] with sincere
admiration."*
—Albert Einstein.
*The Hartford Courant called the same book, "The most significant work on
psychoanalysis since Freud's* Interpretation of Dreams, *[a] masterwork."
And the Atlantic Monthly reported, "Dr. Reik's book is the first to turn
psychoanalysis inside out . . . to study the analytic process from the
standpoint of the inner experience of the analyst."*

The closest I got to Sigmund Freud was his friend and disciple
Theodor Reik. As Reik opened the door of his upper West Side
Manhattan apartment, a neighbor opened hers and screamed at him
in frenzied incoherence. He nodded to her in a friendly fashion as if
he was completely deaf or immune to shock. Then he beckoned to me
to precede him into his office and calmly closed the door behind us.

That's one way of handling problems, I thought. Either the
screams stopped or the room was soundproof. He made no mention
of the hysterical attack as he sat at his desk, lit a cigarette and nodded, through a cloud of smoke, to the facing chair.

If not a facsimile of Freud, Reik was close, and he had worked at it. You could judge the comparison by the 50 photos of Freud on the walls of Reik's office, a shrine in fact, with statuettes on the desk and Freud's books in the shelves.

Reik told me he had lost both parents while in his early twenties and found in Freud a beloved father figure. While studying at the University of Vienna, intending to be a physician, Reik had written for his doctoral thesis the first psychoanalytical interpretation of a literary work—Flaubert's *Temptation of Saint Anthony*, and sent it to Freud for his comments. Freud, even more of an expert on the book than Reik, invited him to discuss it and eventually persuaded the young man to drop his plan to become a doctor and to switch to psychoanalysis. Freud then generously funded Reik's studies and his early struggling years in practice, and gave him a prize for the best scientific work in applied psychoanalysis—Reik's paper on *The Puberty Rites of Savages.*

A year before Hitler's troops invaded Poland in 1938, Reik left for the United States speechless with emotion when saying goodbye to Freud, who comforted him with an arm on his shoulder and a reassuring, "People need not be glued together when they belong together." Freud left for England the following year. Of Reik's two brothers who remained in Vienna, one was murdered by the Nazis and the other committed suicide.

I learned from Reik, who chain-smoked cigarettes he never finished—"I'm not addicted, I've given up dozens of times"—that Freud analyzed the composer Mahler who had a Madonna complex, that Reik had been in love with Mahler's wife though aware he was really in love with Mahler himself [it began to sound like a comic opera], that Neitzsche [presumably when he was mad] believed women had razor blades in their vaginas, and that Freud became celibate at 45.

Specializing in obsessional neurosis, from which he himself had suffered, Reik's first patient in America was a debutante. Her agonizing dilemma was whether to go to the Stork Club or the 21 Club. Reik also treated Lord Louis Namier, a British Zionist leader who was also a Nobel prize-winning physicist.

Hurt by mean-spirited rivals who called him disloyal for daring to suggest not all Freud's ideas were fruitful, Reik had countered with, "Freud himself told me, 'I am not a Freudian.' " Even if this was typical of Freud's self-deprecating humor which Reik thought not sufficiently recognized, it got Reik off the hook on which his critics had hoped at least to harass him.

Reik illustrated this brand of Freudian humor by what sounds apocryphal: Freud was doing the rounds of a mental hospital with doctors who scorned psychoanalysis. When he proposed to show them his method was superior to theirs, they accepted the challenge. Freud then indicated a patient they had diagnosed as psychotic and said he would demonstrate that the man was quite rational. "Who are these men with me?" he asked the patient. "Doctors," the patient replied. "And who am I?" Freud asked. "An idiot," said the man. "You see," Freud concluded triumphantly. "I told you he wasn't crazy."

When I asked Reik, "How would you begin if I was your patient facing my first psychoanalytical session?" he picked up a ballpoint pen and said, "This is a penis." He pointed to one end. "And this is the sensitive part." Never before having associated pen and penis, I changed the subject and asked if Freud would approve of today's psychoanalytical practises.

Reik: He was very much misunderstood, especially in America. He never tried, for example, to apply analysis to schizophrenics which was done all over America. People have misinterpreted even his basic thinking. He would be horrified by the emphasis on sex, for example. It is too narrow, too exclusive. Sex meant for Freud also friendship and affection.

Why is psychoanalysis especially popular in the United States?

Reik: This is very easy. After a short period in which analysis was rejected, American psychiatrists got into a Freud-mania and went overboard. A reaction formation then occurred, too much on the other side. After they reject it, they accept it. Freud reacted with anger to my coming to America, you know. He didn't want me to come. Don't forget that after World War I the worst Americans came to Vienna—many to train as psychoanalysts—and Freud couldn't understand them. He said they mumbled and spoke from the corner of the mouth. The first American patient I treated was an obsession case. All other English-speaking analysts in Vienna were fully occupied after the war, so Freud sent me to this man with an obsession and I treated him for several weeks. Then I gave Freud my first interpretation of what the man's symptoms meant. And Freud said, "Tell that to the Marines!" And I thought for a moment I'd gone crazy, or Freud had gone crazy. I thought, Oh, god, what has that to do with the Navy? Tell that to the Marines! Later of course, I knew what Freud meant. Much later I read that during the war the Japanese gave better treatment to those prisoners who spoke propaganda to their fellow Americans over the radio. One American officer agreed to do it and he said over the radio: "We have been asked to speak to our brothers and fathers who are serving in the United States army to tell them that the Japanese cannot be conquered. And I want also to—tell it to the marines!" ["Tell it to the Marines!" is a British expression meaning, "I don't believe it."]

Was Freud outgoing or morose?

Reik: You'd call him a regular guy, simple, easygoing. He played cards every week. He said once to Ernest Jones, "I understand how you became a lawyer but how you passed your high school exam I don't know, you play cards so badly." He liked music, especially Mahler, and so did I. He appreciated beautiful women, especially the spirited Princess Marie Bonaparte [who used her influence to help move him from Austria to England before the Nazis could arrest him for being a Jew], but he was a highly moral man with a puritanical streak. Like many great men he was a masochist. He was a good husband and a great father.

What was his wife like?

Reik: A wonderful wife and excellent housewife. But she could never understand why her husband was famous. Her fuzzy understanding of psychoanalysis was indicated by her comment about hysteria, when she said, "We had troubles when we were young, too, but then comes the menopause and it's all over!" [This might, of course, have been her dry sense of humor, because Reik also recalled her saying, "I get along well with my daughters-in-law: I never see them."]

And his mother?

Reik: The film about Freud directed by John Huston was quite good, but not as regards Freud's mother. She didn't speak German. She spoke Yiddish. I'll give you a characteristic trait of Freud. At this time he didn't treat his students in analysis, but every analyst had to go through analysis. So he sent me to Berlin to Dr. Karl Abraham for analysis—for nothing. I didn't pay one cent. Not only didn't I pay one cent, I had no means. My father died early and Freud gave me $100 a month. Not only that, he was angry when I didn't want any more money.

Freud was a member of the British Society for Psychical research.

Reik: Yes, but he didn't believe in spiritualism and psychic phenomenon, though it might have intrigued him. [Freud's daughter, Anna, told me that he briefly investigated telepathy and believed it to be possible.] I'll tell you what he believed. When they made Freud Honorary Professor at the University, he thanked them and said, "I am an atheist. I have no national bias. I have scarcely any knowledge of Jewish history. I cannot read Hebrew. And when you ask me: 'What is still Jewish in you?' I reply, 'The most essential part of my personality.' " That's one of those tricks like O. Henry's. [The short-story writer who gave his tales surprise endings] Freud meant the unconscious continuing experience of his people.

Did Freud give you students of psychoanalysis any guidelines?

Reik: Freud said to us, his intimate circle of students: "Never give advice to your patients except in the form of the Ten Commandments. *Don't* do that! *Don't* do that! Patients should be protected by this, especially against self-damaging tendencies. Freud would say, "*Don't* marry the girl now. Postpone it, postpone it," until with new insight the patient can make a good choice. During analysis the patient should avoid making a vital decision, like a change of profession or getting married.

Psychoanalysis wasn't always Freud's greatest interest.

Reik: Oh, no. You see that picture of Freud [on the wall] surrounded by forms of ancient Egypt? [figurines] Freud was once much more interested in Egyptology than in analysis. I'll give you an instance. There came a Mr. Howard Carter to Freud, the man who with Lord Carnarvon discovered Tutankhamen. And Freud sent Carter away to another doctor in Budapest. Freud said, "I couldn't analyze him. I was so interested in what he told me about the excavations that I had to send him to someone who wasn't so interested in archaeology." I would like to have been an archaeologist if I hadn't been a psychoanalyst. I wrote a few books about Jewish pre-history 5,000 years ago, about the mystery of Mount Sinai. It is quite different from what people think it is. [I regret I didn't pursue the subject.]

What was Freud's attitude to capital punishment?

Reik: He was strongly against it and so am I. It is barbaric. It doesn't prevent more murders. Also suicidal people might kill hoping they will be executed—not having the courage to kill themselves. Capital punishment might even suggest to unbalanced minds the idea of murdering someone.

What did Freud think of you?

Reik: He wrote very early that I was one of his best students. Then he said I didn't fulfill his hope. Perhaps I did a little, after he died. But before that, no. It makes me sad that during his lifetime he did not know of my achievements.

Is man basically good or evil?

Reik: I would say there is as much good as evil in man.

Do you believe in life after death?

Reik: No. Young people are afraid of death. Old people, me for instance, are not afraid of death, of not existing any more. But they are afraid of the death struggle. That I am afraid of, of dying—not of death.

Do men understand women better than women understand men?

Reik: It is very difficult for a man to understand a woman. What man can know how it feels to bear a child? But it takes no great imagination for a woman to know what it is like for a man working in an office.

What do you most admire in people?

Reik: Moral courage. The courage to understand that one has in oneself malice, envy, hostility, as have all people.

Has your knowledge of psychoanalysis helped you personally?

Reik: For my personal life, yes. Don't forget, I was neurotic, too. When I went into analysis with Dr. Abraham in Berlin he got rid of my neurotic symptoms. Freud is quite right when he says: "The main thing in analysis is"—this was one of his pessimistic moments—"to change a neurotic symptom into a common human misery." I had a neurosis that I thought I would die at 68, because my father died at that age. I am in two or three days 79. When I was 18, my father had a heart attack and the doctor sent me to a pharmacy for a camphor injection. I ran through the streets and I ran and ran, but I couldn't go on. I had to stand still. Then I came home with the camphor injection and my father was dead. And that filled me with great guilt feelings.

Many think of psychoanalysis as being a long drawn-out affair. Do you know of any astonishingly quick cures?

Reik: Yes, by Freud's daughter, Anna. Her patient was a 7-year-old boy with an obsession symptom. He would walk two steps forward and one step back. Anna was trying to cure him for a long time and was baffled. Then, one day she hit on the cause. "You know, Michael," she said to the boy, "your father doesn't love your mother any more. Perhaps they will go away from one another." And what she said was right. As a matter of fact, the boy's father married another woman. The day after Anna spoke to the boy about his parents his obsession symptom had gone. He walked normally. Bringing his fears to consciousness cured him in one day.

How has psychoanalysis helped you to raise your own children?

Reik: It has helped me be more sympathetic and understanding. I'll give you an instance. My wife died in 1959 of multiple sclerosis. I have two daughters, one is a lawyer and the other a professor in Philadelphia. My wife was many years in a home. Multiple sclerosis is incurable, and she couldn't speak any more. My daughters went to visit her as often as they could but I went only rarely. My daughter Miriam was in analysis, and I was in a gloomy mood and went to see her. I sat down and said, "I wasn't nice to mother. I went to see her rarely because she couldn't talk any more. I couldn't understand what she said, and I had to work to pay the expenses for the home.

When I went there I stayed only a short while, because I couldn't understand what she said." I said, self-accusingly, "I wasn't treating her well: I scolded her sometimes unjustly." Then I had coffee with my daughter. When I opened the door to leave, Miriam was in the corridor and she said, "Daddy, one has to forgive oneself, too." It was very charming. And that is the lesson of psychoanalysis.

As I left Reik's apartment his neighbor again opened her door, but this time just stood there, a silent witness.

Chapter Sixteen. Dr. Murray Sherman

"Psychoanalysis produces change through the analytical relationship; it
does not simply eliminate symptom or complaints."
—Murray Sherman
"I think that unconscious homosexuality is a significant aspect
of paranoid conditions."
—Murray Sherman

Sigmund Freud taught Theodor Reik. Reik taught Murray
Sherman. What has been passed on? What has Sherman to say of his
mentors? And how flourishing is psychoanalysis today, 1995? I asked
Sherman, now in the private practice of psychoanalysis and marital
therapy. He has a bachelor's degree in psychology from Wayne State
University in Detroit and a Ph.D. in psychology from Columbia. He
was managing editor of *The Psychoanalytic Review* for eight years and
is still on the editorial board. His decision to become a psychoanalyst grew out of his therapeutic efforts to treat mental patients and
from his own analysis. "I have always had a persistent curiosity about
how the mind works and what drives peoples' behavior," says
Sherman. "More basically, my interest in human psychology developed from my earliest efforts to understand my mother's inconsistent behavior towards me."

In recent years psychoanalysis has been under attack for being unscientific and out of reach of all but the rich because of the spiraling fees and lengthy treatment. Although, as Robert Stewart points out, "The two world wars . . . played a significant role in the spread of psychoanalysis in the United States: those Army psychiatrists who understood psychoanalytic concepts—conflict, anxiety, defense, dream formation, catharsis—demonstrated more skill than anyone else in rehabilitating soldiers, prodding shellshocked young men to say what was on their minds during brief interventions in the combat zones."[1]

I told Murray Sherman: Theodor Reik seemed to believe what was true of himself was true of mankind generally. He says, for example, that a man who is interested in the decor or ambiance of a room obviously has feminine qualities. And that he personally wasn't in the least interested in the appearance of a room. Now I'm very interested in the ambiance of a room [Sherman chuckled] and I don't think I've got a pennyworth of femininity in me.

Sherman laughed: It wouldn't be so terrible!

I realize that every male has some feminine aspects, and every woman, masculine. But he implies that it's more than average.

Sherman: Reik was extremely obsessive about the concept of latent homosexuality. He did have some extreme ideas. For example, he felt that any man who paid attention to the way he dressed and was careful about selecting his clothes, was exhibiting homosexual trends.

There goes Sinatra!

Sherman: Well, lots of people besides Sinatra. But Reik was extremely careful to be sloppy, so no one would accuse him of that.

That's funny.

Sherman: At one point he was employed as a consultant at a mental hospital, and he went around in such disarray that people were shocked and took him for a mental patient.

If Reik had been your patient, would you have assumed he was a latent homosexual?

Sherman: We're all latently homosexual. That's part of the human disposition.

Then are all homosexuals latent heterosexuals?

Sherman: Yes, I would say so.

But wouldn't Reik have known that?

Sherman: I think there was a particular conflict of that nature. I think, theoretically, Reik knew it. If you were to have asked him, "Aren't all men latently homosexual?" he would have given you a sensible answer. But it didn't affect his behavior.

Is paranoia, as Reik said, generally an unconscious fight against homosexuality?

Sherman: I personally think that unconscious homosexuality is a significant aspect of paranoid conditions, but that view tends to be in disrepute today. Still, I agree with Freud and Reik about unconscious homosexuality and paranoia.

Do you find yourself echoing Reikian ideas to your students?

Sherman: Today I'm more influenced by my relationship with Hyman Spotnitz, well known here in New York City, a man who is well into his 80s and still working as a psychoanalyst. Among those who are with him, he has an excellent reputation, to put it very mildly.

Who are the Freuds, Reiks, and Karen Horneys of today? Would Spotnitz be considered among the leaders of today?

Sherman: In my professional circle he'd not only be a leader, he'd be *the* leader.

Your circle?

Sherman: Modern psychoanalysis. That's the term Spotnitz applied to his system of working with patients.

What are the main things you've learned from him?

Sherman: The primary importance of getting patients to put all their thoughts and feelings into words. Freud said that patients should "say everything," but he gave equal importance to interpretation, explaining to the patient the meaning of his or her symptoms. Saying everything is the shibboleth of Spotnitz's followers. It's far more important to get the patient to express feelings and thoughts than to voice your explanations of what's been said. Investigate what's been said rather than explain it, that's the idea. In time the patient gets new understanding and the problem resolves itself. That and the idea of proceeding with the analysis at the patient's pace rather than a predetermined schedule defines modern psychoanalysis. It includes interpretation but only sparingly.

Are obsessions, compulsions, and phobias the main symptoms you treat?

Sherman: No. Today we treat more general problems: character neuroses; maybe an obsessive-compulsive character; borderline conditions that verge on psychosis. Marital problems are very common. The majority of the problems treated by analysts today are, basically, family and/or marital problems.

Reik complained about his own unsatisfactory love life.

Sherman: [chuckled] I don't think it was more unsatisfactory than average, frankly. Love life is frequently unsatisfactory. It's one of the cornerstones of adjustment to marriage.

In *The Need to Be Loved,* Reik mentions a patient suffering from erotomania who suspected Reik of wanting to kill him, and the psychoanlytical treatment had to be broken off. Why would the patient not have left in a hurry if he suspected Reik of murderous intentions?

Sherman: An excellent question. I heard about the case from Reik. It was a transference reaction and even though such paranoid reactions sound dreadful, their meaning depends upon the way that kind of accusation is communicated. It could actually be an erotic wish in disguise, as when a patient might feel the analyst is going to castrate him in order to treat him like a woman. It has a sort of double meaning in a sexual sense.

You mean the patient wanted to kill Reik, so he thought Reik wanted to castrate him?

Sherman: If that's the patient I think it was, castration anxiety of a psychotic nature was involved. The patient became very paranoid. He thought men were coming over the roof to castrate him, and he thought Reik was part of the plot.

How would Reik get rid of him?

Sherman: He may have just abruptly terminated treatment.

But wouldn't a paranoid patient insist on seeing him?

Sherman: Not necessarily.

Reik never advised you in seminars or in private how to handle dangerous situations?

Sherman: Reik did give advice when those situations arose. Actually there are patients who are difficult to terminate even aside from that kind of extreme. I haven't found it a very common problem. If you're determined to rid yourself of a patient, it's not that difficult. You can simply say things the patient doesn't like to hear and the patient usually will stop coming. That was undoubtedly a case that Reik mishandled and he interpreted things that were better left uninterpreted at the time, and that led to a severe regression with psychotic episodes. So, in that case, I think it was due to Reik's incompetence.

Should he have realized from the man's reactions that he was going into dangerous territory?

Sherman: I believe so. It was very early in Reik's career. He was being supervised by Freud, and possibly Freud himself may have been mistaken. Freud's therapeutic efficacy has also been questioned in some cases.

How did Reik influence you?

Sherman: Reik placed a great deal of emphasis upon the influence of the unconscious and how the unconscious is communicated through very small nuances of behavior. And he encouraged analysts to rely on their own unconscious reactions when they were with patients. In those ways he was an excellent teacher. At a seminar, we would present cases and Reik would say, "Do you know what I would say to such a patient?" and then he would say something like, "When you were younger you must have had a great envy of your sister." And everybody sitting there would get the feeling that that was exactly the right thing to say. We all envied Reik's gift for feeling and communicating the unconscious.

How do you communicate the unconscious?

Sherman: Poets sometimes come close to it and sometimes in a slip of the tongue or, symbolically, in dreams.

This is what Reik wrote: "If Freud could see what was happening in the year 2000," which is pretty soon, "he would be as deeply shaken by a psychoanalytical session in New York as the rabbi of Nazareth would if he were present at a Roman Catholic mass."

Sherman: [chuckled] I'm rather shaken myself.

Because?

Sherman: Psychoanalysis has become very stereotyped, that is, restricted by rather rigid and abstract derivatives of Freud's written work. I have a rather individualistic view of psychoanalysis. What I say about psychoanalysis is not typical. In my own view, psychoanalysis has gone downhill in the last 50 years, mainly because it's been overly influenced by the medical model of treatment. When a person consults a physician he presents a symptom, such as headache, and he expects to be given a medication that will eradicate the pain. Usually, the physician will prescribe a medicine which accomplishes just that. In psychoanalysis, many analysts believe it's up to them to provide some sort of utterance, usually in the form of an explanation or interpretation, that will make clear to the patient why he or she had his or her particular emotional conflict, and thus truly *eliminate* that problem. Psychoanalysis produces *change* through the analytic relationship; it does not simply eliminate symptoms or complaints.

If not as patients, how should they be treated?

Sherman: With a much more personal approach and less interpretation. The combination of a medical model and excessive emphasis on explaining things to the patient has deprived psychoanalysis of much vitality. The patient's own view of how he or she wants to be treated should be given far more emphasis than has been the case in past years. For example, if someone wants to come for therapy just once every other week, that's fine. Even today, in some places, unless a patient consents to several sessions a week, he or she is considered only for therapy, not for analysis, and thus assigned a lower rung on the psychoanalytic ladder. Today, candidates for three-times-a-week analysis are so few that many analysts accept whatever is available. I think that the analyst has to become attuned to responding to whatever a particular patient needs.

Act more like a friend?

Sherman: I don't think an analyst should be a friend. He may be friendly, although some patients feel more secure in a distinctly neutral atmosphere. Personal and emotional factors in the psychoanalytic relationship are recognized but not often written about. I think analysts more or less intuitively realize they're dealing with a personal relationship and respond accordingly. But the analyst's own emotional responses usually go unmentioned in published work.

Do you keep out of the patient's eyeline?

Sherman: Not as much as I used to. With about half of my patients. I sit facing them.

Ever take notes?

Sherman: Not during a session.

As Reik recommended.

Sherman: Yes. Freud did not, and most analysts do not. Not during the session, but they may do immediately after.

Do you agree with Reik that you can't treat a fully developed psychosis with psychoanalysis.

Sherman: It depends upon the severity of the psychosis. If a person is in a catatonic rage or stupor, the first thing you have to do is get into contact with him. You can't do that by explaining the rage or stupor analytically. In their excitement catatonics may literally exhaust themselves to death, unless they're medicated. It is also clinically feasible, by staying with such patients on a 24-hour-a-day basis, and communicating with them in a meaningful way, to eliminate the psychotic episode. [A practice, apparently of the iconoclastic Scottish psychiatrist, R.D. Laing.] But not many therapists are ready to do that, and therefore medication is a more practical alternative. Such remissions, by medication or by psychotherapy, are often temporary unless part of a continuing relationship. When a suicidal risk is truly imminent, psychoanalysis as such is not feasible. But very often the mere mention of suicide by a patient so alarms many therapists that they immediately recommend a psychiatric consultation. Frequently such a recommendation is made more to protect the therapist, or the agency where the therapist is working, than to help the patient. The therapist's own anxiety, or the risk of being sued in case of an actual suicide, often interfere with effective treatment. Some psychotic conditions are amenable to psychotherapy—not to the classical forms of analysis but to analytically-oriented treatment.

Have you succeeded with psychotics?

Sherman: Yes. There are cases I've seen that have been judged psychotic. Other therapists might have thought that hospitalization was needed where I did not think so, and I was able to treat the patient outside of a hospital. If this can be done, it is a true boon for the patient because nothing is more damaging to someone than being forcibly put into a mental institution when it is not truly needed.

What was his or her psychosis?

Sherman: I've been successful with some paranoid conditions and also with extreme anxiety that verged on psychotic development. I believe that extreme anxiety is often at the root of psychosis, even where biological factors cause or magnify the anxiety. This is also the case in many physical illnesses, and if the therapist is able to reduce the anxiety, the physical symptoms may vanish. But the psychosis or illness may return if the patient leaves analysis prematurely.

Do you agree with Reik that depression is rage against someone else turned against oneself?

Sherman: That's pretty well established.

But surely there are other causes of depression.

Sherman: I think you'll find rage directed against the self in every depression, whether biologically-based or stemming from emotional causes. Actually, I believe that the distinction between biological and emotional causes is, fundamentally, an artifact, two ways of regarding a single phenomenon.

A rage against someone else is responsible?

Sherman: Yes. Someone close.

And it's the same for suicide?

Sherman: Yes, directed against that other person.

But then there's suicide for unbearable pain.

Sherman: That might be an exceptional case but, at least in theory, even in such cases a direct expression of rage might alleviate the pain.

Do you stick to Reik's view, which I believe he got from Freud, that you should tell patients what not to do, like the Ten Commandments, but never what to do?

Sherman: [chuckled] Never say never. I seldom tell patients what not to do, and I rarely tell them what to do, unless I feel it indicated in that particular situation.

What writers do you like?

Sherman: I used to like Norman Mailer until he became rather stylized and began to copy himself. I like the plays of Harold Pinter.

Would you like to psychoanalyze the writers?

Sherman: I would. But I'd be interested certainly just speaking to them, aside from any analytical contact.

But don't you think the subject of their work would make psychoanalyzing them an interesting exercise? Mailer also stabbed his wife. And Pinter's plays are so strange.

Sherman: I believe Mailer's stabbing his wife might have been (done) under the influence of drugs; that, plus a need to adopt a kind of supermacho stance in life. I don't find Pinter strange. I find him very meaningful.

You don't think his work is like *Waiting for Godot*?

Sherman: I don't find Beckett strange either. I can see where people might think such writings strange if they don't accept the dissociated nature of mind, of which most people aren't particularly aware. Pinter capitalizes upon the remarkable inconsistencies in ordinary people's speech and behavior. We all behave in contradictory ways in different settings, depending upon our specific needs in those situations. Modernist writing in poetry and novels, like the work of T.S. Eliot and Virginia Woolf, veers off abruptly in unexpected tangents, and this reflects the way in which our minds actually function.

You see that reflected in patients perhaps.

Sherman: I see it reflected in everybody, from presidential candidates, to patients, to all of us. I don't find that so strange.

I presume that's different from dissociation.

Sherman: Not so very different. In degree, not in kind. To me it seems quite similar.

Is psychoanalysis a flourishing form of therapy?

Sherman: Less flourishing, partly because there's considerable disappointment with the therapeutic results. Other theories and ideas come along and get the attention of the moment. Another predominant reason is the expense. Insurance companies won't reimburse analysis. They'll reimburse some psychotherapy and they may reimburse brief analysis. But if you're an analyst and you want your patient to be reimbursed by insurance, you're better off calling what you do psychotherapy. Insurance companies, and managed care in particular, look upon the therapeutic efficacy of psychoanalysis with great skepticism.

So do you generally treat wealthy people?

Sherman: Most of my patients would be considered middle class, although a few are wealthy. However, there are many psychoanalytical institutes, including the National Psychological Association for Psychoanalysis, the organization that Theodor Reik founded, which offer low cost analytic therapy for whatever the patient can afford.

Another training institute where I'm affiliated, the Center for Modern Psychoanalytical Studies, also provides psychoanalysis for low fee applicants. Modern psychoanalysis is a form of analytic treatment initiated by Dr. Hyman Spotnitz. Almost all analytic institutes offer low cost treatment, partly in order to train psychoanalytic candidates.

What do you think of the accusation made by some that Freud hadn't the courage to stick with his idea of children being molested by parents? That he lacked moral courage in not maintaining his first thoughts about child molestation?

Sherman: I don't think that's true. I think Freud had a great deal of courage. After all, he discovered and defended psychoanalysis against practically his entire culture. It's a matter of whether or not these accusations of abuse are merely fantasies. What's being discovered today, is that abuse has been suggested by the therapist. I believe that Freud came to recognize that fact and therefore changed his position from actual abuse to fantasies of abuse.

Some hypnotists are accused of inducing fantasies.

Sherman: Yes, I believe the accusation is justified.

Do you still interpret dreams?

Sherman: Sometimes, when it seems indicated.

When would it seem indicated?

Sherman: There are different indications. Some patients are so afraid of direct contact that they present dreams in order to avoid closeness to the analyst. In such instances, I try to make certain that my interpretation will not arouse anxiety. After a patient has been in therapy for several years, my interpretation might be intended to raise the level of anxiety. My own feeling is that understandings reached in direct interactions between patient and analyst are more worthwhile than dream interpretation. But dreams are often helpful in indicating the patient's preoccupations, the relationship with the analyst, and general emotional stability.

How do you like this? The analyst's nightmare: An analyst dreams that he's listening to a patient and he wakes up and it's true.

Sherman: Analysts fall asleep [during sessions]. That's a common phenomenon. Patients can be repetitive. More than that they can go on without communicating feelings or without contacting the analyst, so you get to feel you're superfluous, closed out of the situation, killed off so to speak.

Do they quit when they find you've fallen asleep?

Sherman: Usually no. Sometimes it can be quite salubrious. It shakes things up. It makes a change. Patients realize a different kind of relationship is needed, a new kind of communication seems to be indicated. A lot depends on how the analyst handles it. If the analyst is able to communicate an acceptable meaning to his falling asleep, and without feeling guilty about it, that generally takes care of the situation.

How many patients do you see in a week?

Sherman: Fifteen or so.

Is it exhausting overall?

Sherman: Yes and no. I don't know if exhausting is the right word. It can be tiring or difficult. It can be quite stimulating and exciting and even endearing. A lot depends on whether you have an engaging relationship, whether the two of you are communicating, or one person is doing all the talking.

I thought that in classical psychoanalysis—the so-called "talking cure"—the psychoanalyst hardly said anything, except, "What do you think it means?"

Sherman: That's not true today. It was true many years ago. But I think analysis has, despite my previous statements, worked its way out of that. I don't think analysts are as silent as they once were, but extended silences do occur on the part of both patient and analyst.

Have you any interest in brain research?

Sherman: A great deal. I've read Sperry and also Gazzaniga on the split brain, and Penfield, one of the first people who did actual and very dramatic observations of the brain in action. I'm particularly interested in the relationship between the inability of the split-brain patients to express what's been recorded in one brain hemisphere and most people's inability to give voice to the unconscious.

Is Freud still the master in the psychoanalytical field?

Sherman: I should say so. He's still considered fundamental to everything that's developed since his time.

How about Jung?

Sherman: I like some of his early writing that deals with working directly with patients, but I think Jung was carried away by some of his abstract ideas and his own fantasies. He was certainly not exemplary in his treatment of Sabina Spielrein, a patient and later a psychoanalyst, who apparently was psychotic or close to psychotic when he treated her. Part of the treatment consisted of sexual intercourse. During this treatment Jung wrote to Spielkrein's mother, ". . . you should pay me a fee as suitable recompense for my trouble. In that way you may be *absolutely certain* that I will respect my duty as a doctor under all *circumstances*." [Jung's emphasis.]

And he was providing the sexual intercourse?

Sherman: Yes. There's been a book about it.[2]

In contrast, Freud was puritanical, wasn't he?

Sherman: He was certainly abstemious after he had children. He lost interest in sex at about 50.

Freud's question, "What do women want?" implies that they baffled him. Do you agree?

Sherman: I don't think Freud understood Dora very well—one of his first patients. He later recognized that his treatment was not what it should have been.

What was Freud's mistake? And how would you deal with her today?

Sherman: I think Freud's main mistake was one I've emphasized to you: he plied Dora with interpretations of her dreams when she was mainly trying to avoid contact with him. He explained to Dora that she wanted oral sex with a married man and that was what was causing her cough. She ran away from Freud as soon as she could. This case taught Freud not to repeat such behavior with patients, and today any analyst would be far more respecting of Dora's need for distance. We all have our limitations. Freud's attitude would be classified as chauvinistic today, but that was the norm of the time he lived.[3]

Although as he was investigating women who confided in him, shouldn't he have been more perceptive?

Sherman: We are all limited by the culture around us. Actually, Freud wasn't that astute with men, either. In some ways he was; in some ways, he wasn't. Jones, his biographer, said he was not a good judge of people, that he was overly credulous. But being credulous led Freud to the discovery of psychoanalysis.

How did being credulous help Freud?

Sherman: He took people at their face value. Somehow, when people felt that, they were more expressive and he found out things. If he'd been skeptical, they wouldn't have expressed so much.

You said that your mother's inconsistent attitude partly led to your becoming a psychoanalyst. How was she inconsistent?

Sherman. She was very punitive on one hand. It's a loaded topic. But she also fed me. I could never figure out why, if she was so disappointed in me, she had to punish me in the way she did, but would take the trouble to feed me as well as she did. Later I had difficulty in getting married. Consequently, I would go out with girls for a period of time and usually they were gentile—I'm Jewish—and get to the point of marriage and would not go through with it. That was my major reason for getting analyzed the first time.

And that was a result of your mother's inconsistent behavior?

Sherman: My mother was very concerned that I marry a Jewish girl.

Most Jewish mothers are, aren't they?

Sherman: Yes. She said, "If you marry a shiksa, don't come home." I don't think she would have carried through that threat, but it was enough to get me stuck on shiksas.

Did being analyzed get you to understand your mother's inconsistencies? Did you ever resolve the problem?

Sherman: I became more accepting of it in more or less recent years. I'm still in an analytical relationship, incidentally.

You go to a psychoanalyst?

Sherman: Yes.

Is it the great one? [Spotnitz.]

Sherman: Yes. I doubt that problems like that are resolved, as I said in one of my previous comments. Symptoms don't disappear. They take different forms. A person changes. You come to an understanding. You come to an acceptance.

Did you come to an understanding?

Sherman: That in itself represents an understanding, the fact that my attitude towards my mother was the result of being confronted by her inconsistencies.

Were you ever able as an adult to face her with it?

Sherman: I never faced her with the fact she was inconsistent. I used to get into fights with her and arguments. Quarrels. We were at odds. Mostly before I was married. Not after I was married.

Was it physical punishment?

Sherman: Yes.

What did your father do?

Sherman: When my father was home he was protective. She wouldn't hit me when he was around.

She's dead now?

Sherman: Yes, for a number of years.

If one has resentments toward a parent like that, not getting enough encouragement . . .

Sherman: She encouraged me. She was very eager that I get educated . . .

Right. And to marry a Jewish girl and be a success. But if you do have resentments, as you did, should you have tried to confront your mother with it when you were older and say, "Why were you like that?" Does that help?

Sherman: I did confront my mother mainly in terms of her being punitive. Told her she hadn't treated me well. What she said was, "When we have children . . . When you have children you'll see what kind of parent you'll make." I slipped there, incidentally.

What was the slip?

Sherman: I said, "When we have children."

Oh, I see. The Oedipus complex. Did you have any children?

Sherman: No. So she could never tell.

How does modern psychoanalysis deal with the situation?

Sherman: It's the verbal expression that's important, the putting things into words. You may have certain thoughts, like I had certain thoughts about my mother. But when you asked me about it, I developed feelings about my mother I wouldn't have had. Even though I don't know my mother any better than when I spoke to you, but it becomes a different situation when you express things in words.

When you get up in the morning to start your day's work . . .

Sherman: Are you thinking of becoming an analyst? I can give you all the encouragement you want.

Do I sound like one?

Sherman: You're getting there.

Do you face the work day enthusiastically?

Sherman: It isn't 100 percent enthusiasm. I sometimes wish I didn't have to do it. It's challenging. You always find out things. It's a process of discovery. You're confronted with challenges. How are you going to present something? How are you going to respond to something?

Do you feel that the mental well-being of your patients depends on you?

Sherman: Not 100 percent. Far from that. But to a certain extent—definitely.

Chapter Seventeen.
Henry Murray

*"He [Henry Murray] was undoubtedly one of the outstanding
psychologists of his time. He was one of the small number of leading
theorists in the study of personality, and his work was known worldwide."*
—Robert Holt, professor of psychology, New York University

Henry Murray walked slowly into the living room of his spacious Cambridge, Massachusetts home, wearing a dressing gown and supported by a cane. Though he had suffered two strokes, was in his late 80s and tired easily, he seemed eager to talk of his experiences as one of America's most influential psychologists. While director of the Harvard Psychological Clinic during World War II he accepted the challenge to put his knowledge of personality to a critical test: deciding which OSS recruits were most suitable for dangerous and tricky assignments. The OSS was America's wartime intelligence service, the forerunner of the CIA. He had the recruits live together in small groups and gave them a battery of tests—both under stress and under relaxed conditions. For one test Murray had a recruit try to assemble a five-foot cube from wooden poles, blocks, and pegs. The man was told he would need two assistants to help finish the job in the prescribed ten minutes. In fact, those helpers were psychologists on Murray's staff. One of them played a helpless, passive role; the other delayed the task with stupid suggestions and by criticizing and needling the recruit. Masquerading as helpers they were so effective that never once in thousands of attempts did a recruit complete the task. That, of course, wasn't the purpose. The scheme was a setup by Murray to see how the men responded to stress.

Murray also used his own thematic apperception test [TAT], still the most used projective test after the Rorschach-inkblot test. To estimate the strength and prominence of an individual's needs, Murray produced cards showing people in ambiguous situations, and asked whoever was being tested to invent a story about each picture. Murray assumed that the subject would project himself into each scene, and when expressing the protagonist's needs, would be expressing his own. Among other questions I wanted to ask Murray was how well the OSS recruits he had okayed for hazardous missions lived up to expectations.

He greeted me with a smile, carefully lowered himself into an armchair, nodded for me to sit opposite him, made sure my tape-recorder was working, and began.

Murray: I'm starting off with optimism in the hope the words will come as I go along, but I haven't recollected events involved here for a long, long time. Let's see.

How did you become a psychologist?

Murray: I was working in chemistry at the Rockefeller Institute in New York. One of the men here at Harvard, Professor Henderson, wanted me to shift to psychology. He had heard me say it came more naturally to me than chemistry. So I got an invitation to dine with the Harvard faculty of philosophy and psychology. At that time [the 1920s] philosophy and psychology were one. Psychology was only represented by two people, and the philosophers had about nine. Alfred North Whitehead was the best known. There was William McDougall, the psychologist, and altogether about nine of us after dinner sitting in a room where the chairs were arranged in a circle. I started talking to the man on my right, Perry, and then noticed a man leaning over to engage me in conversation. When I looked away, there was someone else sitting in his place. It was an attempt to let everybody get a crack at me, to vote afterwards—don't take him, or take him. McDougall was next. He said, "I'm going to show you something. Say what you think it is." He threw it on the table. This was no trick for me. I'd done a lot of dissecting of all kinds. I was very interested in anatomy at one time. I took a chance. I could have said, "That's flesh." Instead, I said, "That's a rabbit's lung." I think it was a rabbit's liver. McDougall said, "That came out of Margery Crandon [a famous but fraudulent Boston medium*]. It's ectoplasm that she produced." So I felt I was hanging by a thread, and if I said anything else I might be excommunicated.

McDougall and Harry Houdini were on a committee to investigate Margery Crandon and went to experts like you, for advice. Hudson Hoagland, also a scientific investigator, told me that he called Margery Crandon a fraud to her face and asked her outright why she did it. According to him she replied: "Why does Mary Baker Eddy?" [the founder of Christian Science], meaning, I suppose, that as a famous medium she was enjoying worldwide attention and acclaim with professors and clergymen hanging on her every word. How can a woman who is reasonably religious, as she was apparently, use her dead brother whom she had loved, as a fake communicant? Doesn't she have to be perverted to do such a thing?

Murray: We're all a little perverted. I've got used to things like that. You see so much of it in psychological practice.

Isn't it slightly insane?

Murray: No. We had a word to describe what would the compass be of a sphere in which you would say, "He's relatively normal?" And if he got outside that, "He's definitely pathological or abnormal." Some people have a huge area where they're more or less normal. Most people have a rather tight area, rather smaller. They don't like to call anybody abnormal or pathological, unless those people have definite delusions and so forth. I lived only a couple of blocks away from

---

* During her seances Margery Crandon purported to be producing "ectoplasm" which emerged apparently from her mouth and ears, and even at times from her vagina. In a similar bizarre fashion she claimed to have materialized the hand of her dead brother. McDougall, who attended many of her seances believed the hand had been faked from animal lung tissue. And Dr. W.B. Cannon, a Harvard professor of physiology, and Dr. H.W. Rand, associate professor of zoology, after examining the "ectoplasm," agreed with McDougall. Margery's husband, a surgeon, was suspected of supplying the material for this extended hoax. She was eventually exposed as a fraud by Harry Houdini, the magician, who knew a trick when he saw one.

Margery Crandon in Boston and went to several of her sessions. Morton Prince [a noted psychologist] also held seances and demonstrations of hypnosis. He once invited some people, myself included, around to his house after dinner. There were eight men and one or two women. Prince then hypnotized a man who lost his sensory perceptions. He had anesthesia, so you could stick him with a pin and he wouldn't feel it. Then Prince said, "Turn out all the lights and we'll sit around this table and hold hands. Now be sure you've got the right hand of your next-door neighbor, and don't let go." And so forth. The hands were under the table. I had my eyes very much wide open, and all of a sudden I got a book in the eye that took me six weeks to recover from. The surface over the cornea was damaged. Dr. Walter Cannon's wife, Cornelia—Cannon coined the term homeostasis—was sitting next to me. She said that the man just beyond her on the other side, instead of holding her right hand, was holding on to her left.

A well-known trick of mediums.

Murray: Oh, is it? So he had a free hand to throw the book [and pretend it was done by a spirit or poltergeist].

What did you think of Margery Crandon?

Murray: She was exhibiting in her home a conversation between her and her dead brother. I don't think there was anything genuine or supernatural about it, and it wasn't anywhere near as exciting as a lot of adventures I've had myself.

[Biographer Forrest G. Robinson later detailed some of Murray's "adventures" which made me understand why he considered Margery Crandon's activities tame. Robinson revealed that Murray, who was married to a conventional Bostonian who thought Carl Jung was "a dirty old man," became involved in a wild affair with Christiana Morgan, also married, the beautiful, sadomasochistic daughter of a Harvard pathology professor. Christiana thought Jung was a genius, and to develop what he recommended—"the active imagination" or her "visions"—she encouraged Murray to join her in weird sexual rituals. They consummated their affair in an apartment near Harvard Yard, where Murray once wore a dress to bring out his "feminine dimension." At times they engaged in mutual flagellation and drank drops of each other's blood.[1]

Christiana Morgan's biographer, Claire Douglas, gives a more sympathetic portrait of her as a talented and passionate woman who made an important contribution to the early development of psychoanalysis but who died "unrecognized and, in the end, unloved." She followed Jung's advice to tap her unconscious and he used the resulting "visions" as the focus of a series of seminars. According to Douglas, while involved with Murray, Christiana had affairs with Chaim Weizmann, later Israel's first president, Lewis Mumford, and philosopher Alfred North Whitehead.[2] Even with those distractions, she became so jealous of Murray's absorption in his work as director of Harvard Psychological Clinic, he feared she might go mad. Instead, Christiana began to drink heavily. It was at the height of his affair with Christiana during World War II that Murray was chosen to test recruits for clandestine operations.]

How did you select members for the OSS?

Murray: We examined them for four days. You could tell ahead of time that it wouldn't be very good, because we didn't know where the men were going to be located and with whom. Some were sent on missions that didn't fit their skills. We'd examine a fellow and say, "This is a fine guy who can do all kinds of things." Then they'd send him to Indonesia where he couldn't understand the language, or didn't like someone he was working with. The OSS people worked in fairly small groups and they had to get along with the others or it didn't work well. Someone wouldn't get along with this kind of person or that and we heard he wasn't any good. Then we found out from someone else that he was very good. This was after we had put him through about 38 test situations where we watched his behavior.

I'll tell you something else I did, and I don't think anyone had done such a thing before. I was coming up in the train in the morning from the country to Harvard Medical School, shortly after the Lindbergh child had been kidnapped. Two ladies were sitting in front of me and one said: "I had the strangest dream last night. I saw the Lindbergh baby." The other one said, "That's funny, so did I." Then they told their dreams. They were different, of course, but both had this feature of concern with the Lindbergh baby. I thought: That's a strange thing. If you can find two people offhand dreaming of this, there must be at least two hundred. So, when I got to the clinic where I worked I called up the *Boston Herald* and got them to make a little announcement saying that a man named Stevens was investigating dreams of people on the Lindbergh kidnapping, and invited them to send in their dreams to R.L. Stevens [Robert Louis Stevenson said that his idea for *Dr. Jekyll and Mr. Hyde* came from a dream], and the address that I gave. What would you think I'd get from the five-line announcement?

At least 200 replies.

Murray: I got 1,300[1], several from Australia, from Buenos Aires, and from every state in the Union. Some newspapers picked up the announcement and enlarged it, saying, "Psychologist investigating this has a way of finding the criminal," that sort of thing. I was doing it partly to find out what kind of person would have that kind of dream. One man sent me an oil painting of what he'd seen in a vision. He said he'd had a vision after the sinking of the Lusitania. This was before there were any hints about what had happened to the Lindbergh child. You could measure statistically: was the baby alive or dead? If dead, was he buried? Did he drown? All the different possibilities. Was he imprisoned anywhere? If buried is he in woods with trees—in how many dreams did that come out? Just by chance, probability, I should have got twice as many right as I did. I divided my report into women who had more than five children and those who had less than five. The women with more than five dreamed the baby was not found or was dead. Especially those with one baby dreamed he was found. In several cases they found him and the Lindberghs came and embraced them and said, "Good, you brought us back our child," and things like that.

Were any remarkably accurate?

Murray: No. Not many. It was very disappointing. There were five fairly accurate dreams. By chance alone there should have been seventeen.

Do you study your own dreams?

Murray: I have a lot of dreams and some of them are on the dark side. I'm very keen on my daughter. Erik Erikson and others, in playing me off in some way, would interpret me in terms of addiction for my daughter. There is a question whether I love myself more than I love my daughter. We had a joke at meetings at the clinic where we analyzed each other. I'm not the only one who will tell you I am very keen on my daughter. But I've had quite a few dreams of her being very sick or, in one or two cases, dying or dead. And I'd say, "God, that was just as real as anything!" And I'd call her in the morning— and nothing at all—not even a headache.

How about apparitions?

Murray: Since I've had this stroke I've had some in the mornings before I'm fully awake, as I'm dimly passing from sleep into nonsleep, the hypnosis state. I haven't got my eyes wide open. The whole thing is a sort of London fog, and my whole environment is a little foggy. The last one amazed me because the fellow is a friend of mine. In fact this man, seeing I hadn't published much and my writings were very scattered, thought he'd be a great help to me by bringing them together. I was opposed to it, but when I got this stroke my resistance was low and he went ahead. Maybe I was a little cross with him for what he was doing. The publisher, *Harper's*, wanted to advertise the book and I didn't want any advertising. So I was a little cross. And as I was waking up there he was, just like you are now, sitting down and looking at me. I looked at him and strangely enough he didn't say anything, and I had no impulse to say something. Ordinarily I would have said, "Now what the hell did you do that for?" If I saw him [in reality] I'd be impelled to talk to him right away. We just looked at each other for quite a long time—20 seconds—that's a long time to look and not say anything. Then he disappeared and I went to sleep. They [apparitions] usually speak and I speak back, and my wife wakes me up and says, "What did you say?" And I say, "Oh you're awake are you?" I'm very interested in recalling the conversations and thinking of the details. It's perfectly natural. There's nothing peculiar about that. I had one the day before yesterday. One principle behind it is to keep an open mind for anything that might turn up. I love to believe in these phenomena as being important. I was on the side of awe of Jung. I knew of his interest in them and when he got off on that, a bird would light on a bush and he would say, "Somebody has just died and his soul is in that bird."

Was he serious?

Murray: He believed in the play of his imagination. He was pretty serious about everything. He had a sense of humor all right, but it wasn't prevalent. The first day I saw him I was quite astonished. Later on, I got to know him and spent a three-day weekend with him in Switzerland, sailing down the lake and stopping at his hideout for a couple of nights. I always go to a place where my target [subject for a possible biography or study] lives—walk over the paths they walked over, and climb the mountains they climbed. Jung didn't climb mountains, which disappointed me. I got much less in awe of him in time, but I still enormously admired him. After a while, I would tease him a little bit. He took himself very seriously, though.

Jung wrote about an extraordinary personal experience. A neighbor had recently died and in a vision Jung saw the man again, alive. In his imagination, Jung followed the man to his nearby house where Jung saw specific books in a bookcase. Next day, in reality and for the first time, Jung went to the neighbor's house. And the books he saw in the vision, or using his "active imagination" as he called it, were there in reality—all of them in exactly the same order. How do you explain that?

Murray: If your mind is filled with some subject, anything in the environment that's related to it will strike you.

But Jung reported that he saw five red volumes in his vision and they were there in fact. As a psychologist, how would you explain it?

Murray: Morton Prince was interested in problems like this, that came up in his practice. I worked with him in the beginning and he used to say, "If I had two or three weeks and I could investigate the person here I would get the answers. I know how to get the answers, especially if they are hypnotizable. I can find out things in the unconscious. But I can't tell you right away because, in each case, certain unique features fit the case."

Didn't Jung tell you that Freud admitted to having seen an apparition?

Murray: Jung said that Freud told him that once he looked up from his desk and saw an apparition of a dead first cousin or brother standing in the doorway. Jung told me that story to indicate how close Freud's consciousness was to the intrusion of some image that didn't correspond to any reality. I think Jung may have exaggerated because, don't we all once in a while think we've seen someone, perhaps in a crowd, and we look again and it isn't the one we had supposed? Jung was apt to dramatize things and to make a good story out of it.

But I thought Jung believed in the reality of apparitions.

Murray: Oh, yes. My impression was that Jung was very suggestible, and so were many of his students. They were hypnotized by him all the time—thinking in his terms and talking in his terms. He had a tremendous, magnetic effect on people around there. They were all in some way his patients, but he passed them on to other people who were doing therapy. We got on well together. We had things in common, like sailing. We talked as fast as we could. He did most of the talking, I'm glad to say, and I learned a lot from him.

Jung heard ghostly voices and his doorbell ringing when nobody was at the door. If a patient reported such experiences to you would you expect he needed treatment?

Murray: Not necessarily. Not on that basis. Those are fairly common hysterical phenomena and usually they'd occur to a hysterical personality. Although I don't think Jung had that. He was on the psychotic side. Not neurotic, the way Freud was. That was the distinct difference between them.

Does that mean you discount the voices and experiences as objective events?

Murray: Do you know the story Jung told when he was visiting Freud and talking to him about these phenomena, and there was an explosion in Freud's bookcase?

Yes. Two explosions, in fact. Do you think, as Freud said, that they were due to the wood expanding and the fact that they happened while Jung was talking was just a coincidence?

Murray: I wouldn't be surprised. But Jung was anxious to believe. What started him was when he was graduating from medical school and investigating his cousin, who was a medium. When I mentioned J.B. Rhine's E.S.P experiments at Duke University to Jung, he had read about them and gave them great weight. He thought they were decisive in proving what they set out to do. Jung would start a seminar for two or three weeks and tell about some of his own visions. He would engage in what he called "active imagination," with himself first. He had a lot of theories and undoubtedly those theories are something to do with what he saw, because you can't keep volition out of it entirely. First, you have to have the volition that you're going to have a vision. You get to a comfortable place and cover your eyes and get into a state of mind to call up some image on the inside of your eyelids. Then you follow this vision. It is always open to manipulation on your part. You say, "That isn't interesting. At this point I ought to have something about the sea." So the sea will come up and so forth. I've definite proof that that occurred with Jung and some of his patients. He told them about his experiences and then for a few weeks they were all seeing just what he saw.

Were Jung or Freud ever hypnotized?

Murray: I don't know. That would have been interesting.

Has anyone other than Freud and Jung deliberately delved into their own unconscious minds?

Murray: Herman Melville. I've been writing his life for quite a while. [which he never published] But now my mind doesn't bubble as it used to. I may have gotten to the end of my tether. I'm 88. I sleep quite a lot. That's one of the consequences of the stroke. I sleep several hours a day, but I usually get free from sleep about 4:30 p.m. and I can stay awake for an hour and a half.

Jung got an honorary degree from Harvard. Why not Freud?

Murray: Freud was voted a degree, and got the highest number of votes. He was number one on the list. Then we were told he could not possibly come to Harvard: He had cancer of the jaw and told someone he would never leave Vienna. If he refused the invitation to Harvard [which it seems he did] a committee had to appoint a substitute. We thought they would choose a historian. At the time psychology was a very small discipline, and didn't have a high reputation in the intellectual world. So, to avoid that historian we were given three places to fill with three world-known men. The third man to accept was Piaget [expert on child psychology]. At the time he was hardly known; it turned out to be as good a choice as we could have made.

Didn't Jung also handle dangerous patients?

Murray: He would take people that were pretty damn psychotic on the insane side, but think they were treatable, and communicable, and relatively normal. Other practioners wouldn't have anything to do with them. They were scared. Jung told me he was talking to one of his patients after having seen him about ten times. And this fellow drew a revolver and put it on the table, saying, "What you just told me makes me feel I wouldn't kill you. I came here to kill you, and you used a word that melted my heart." Jung loved such stories. He was very suggestible and didn't realize what a terrific factor suggestibility was in his patients and seminars. He was also a great talker. Carl Binger had been working in the Rockefeller Institute and I met him there when he decided to shift from medical research to the practice of medicine. So he went to study in Freiburg, Germany, and his medical reading had to do with blood pressure and cardiac symptoms. After about a month he found he was getting some of the symptoms he was reading about. And they took his blood pressure and it had gone up. A lot of the physiological measures were occurring in him from what he had been reading, a phenomenon they speak of as The Medical School Syndrome. Binger decided to consult Jung and was there for a month, seeing Jung every other day. When Binger got back, I said, "How did it go?" He said, "I got well. Jung cured me. We just went into the garden and walked around and Jung talked on and on. It was very interesting talk, and I got better." When I visited Jung a year later—I saw him every three or four years—I said, "I hear you saw my friend, Carl Binger." "Yes," he said. "He came here and talked himself out of his illness. I just kept quiet and listened to him. And he talked himself well." That corresponded to exactly what I knew about them, especially Jung who, as a rule, did all the talking. Of course, Jung was a very good listener, too—as a physician has to be. But ordinarily he had so much to say, and the ideas just bubbled up all the time. Carl Binger was no mean talker, and was quite articulate. So they both thought the other person had done all the talking, which interested me. It came at a time when I was making an observation of other people talking. We set up a situation when I was working at the Harvard Psychological Clinic, and had a visitor for lunch on Wednesdays. We'd purposely arrange it so that the visitor would do all the talking. I'd shut up, if I could. Then we'd make a tournament out of it, like a tennis tournament. We'd invite two people that were well-known for their talking and see who would talk the other one out.

What did you think of William McDougall?

Murray: I liked him. He was British and didn't fare very well in this country, I don't know why. He ran into John B. Watson [founder of behaviorism] in the first place. Watson was an American and made the differences in their views of psychology appear as being British and American. And they went for Watson the Behaviorist rather than McDougall the Vitalist. I didn't pay any attention to Watson. He later went into advertising. Forrestal, [before he was U.S. Secretary of the Navy] had a number of college graduates from Princeton, Yale, and Harvard to dinner once a month to discuss something he wanted to talk about. Watson was there once and I sat next to him, and he said, "You know how Freud gets his results? He gets on the couch with his patients. That's all it is. It's just sexuality on the couch." I was a medical student then and he told it to me as if it were a fact. It was more like Watson himself, than Freud.

Forrestal committed suicide from depression among other things.

Murray: Yes. He called me up on the telephone from the hospital, but didn't give any intimation that he was in bad shape. I ought to have gone to see him right away.

Do you believe in electroshock treatment for such cases?

Murray: Yes. We've had some remarkable results. I knew a philosopher named Schaefer who'd go into depression, then he'd get a shock and go right back and give lectures, and be in fine shape. He was about 40 or 50 then. He worked under a man who specialized in giving electroshock treatment. It worked very well. I also knew of an extreme case. An acquaintance started three years ago on a hunger strike and went down to 75 pounds. They said she was right on the line: one more day and she wouldn't recover. They gave her electric shock treatment and she was cured in one shot. They gave her some more afterwards, but all the change had taken place right away. She was a different person, better than she was before the hunger strike started.

Does W.H. Myer, the early psychic researcher, interest you?

Murray: Very much so. One of the founders of the British Society for Psychical Research. I lived in Myers' house in Cambridge, England, for a year. I spent a year at Trinity College. Myers' house was reputed to be haunted and people were interested on that account. Forty or 50 people, including Lady Darwin, called for the first two or three months that we rented the house. They all intimated in one way or another that they felt strange things went on in the house. I heard strange noises there, but I hear strange noises in any house where I sleep. We call that Lady Macbeth. It's supposed to be her footseps walking around at night. But every house has certain creaks in it that make unaccountable noises. We took it as an interesting reputation. Whenever there was the slightest thing we could fit in with that conception, we did. We were told it was in this house that members of the Psychical Research Society met.

Wasn't Myers planning to visit this group a couple of weeks after he died?

Murray: Yes, that was the arrangement. He died in Italy while Dr. Axel Munthe [author of *The Story of San Michele*] was there. And William James was also present, waiting outside Myers' bedroom door, hoping to record his last words and, if possible, any after-death communication. But James' notebook remained blank.

Henry Murray, who died on June 23, 1988, left an after-death communication, in a nonmystical sense. Since 1970 he had been giving a tape-recorded account of his life to author Forrest Robinson, on the understanding that it would be published posthumously, much of it detailing his extraordinary and intense relationship with Christiana Morgan. About the same time, a biography of Christiana appeared. I discussed both with Henry Murray's widow, Nina, also a psychologist. She had been his second wife.

Nina Murray: The biography of Christiana is full of nontruths. Although I didn't know Christiana, I must say the biography made her alive: for the first time ever I felt she was a human being. In that way it was a wonderful biography, but it was so twisted to make it come out as Jungian—the author really distorted the facts. She made Harry [Murray] out to be a woman hater. At Harvard, Harry had more women in his clinic than anybody else. She also said he had taken Christiana's name off the coauthorship of the TAT, and that was a typical male power move. The truth of it was that Christiana herself insisted that her name be taken off because she was sick and tired of getting all the mail. With its publication they got a lot of communications they had to answer—like "Why didn't you do this?" and "Haven't you thought of this?" She didn't want to do any of that, so she begged Harry to take her name off and he finally did. Now he's being accused of being this monstrous male.

The Christiana biography says he came to loathe her.

Nina Murray: I'm sure he had ambivalence, but he didn't loathe her.

She apparently also had an affair with Chaim Weizmann. Where was that?

Nina Murray: In New York.

And reportedly with Alfred North Whitehead and Lewis Mumford.

Nina Murray: Absolute nonsense. She had very close friendships with them. Harry was a very good friend of Whitehead and Mumford and Christiana, and Mumford had a very interesting correspondence which is all in the archives. Harry and Christiana helped arrange Mumford's library and stuff like that, but the implication that they were lovers is nonsense.

How did your husband describe Christiana to you? Did he respect her?

Nina Murray: Very much so.

Do you feel sympathy for her?

Nina Murray: I guess so. But I have very negative ideas about Jung and what he did to them. I think Jung crucified them.

You mean he mixed them up and made their lives miserable because they followed his ideas?

Nina Murray: Right.

But apparently neither blamed Jung.

Nina Murray: No. That's my interpretation.

Your husband survived pretty well, don't you think?

Nina Murray: Uh huh. But mainly because he gave Jung up.

But he told me he visited Jung every couple of years.

Nina Murray: Oh, yes. And he'd see Freud and he knew everybody but in his own work he began to follow his own path.

How do you explain the sadomasochism, the whippings, and exchange of blood? Was Christiana following Jung's teachings to use the active imagination and to induce visions?

Nina Murray: I believe it was a sort of Nietzschian idea. They were great followers of Nietzsche, of understanding all aspects of sex and not to miss any part of it. To be sure to do it all. It was a kind of experiment.

Nietzsche suggested a man should wear a dress?

Nina Murray: I don't know if it was Nietzschian but a sort of experiment in orgiastic stuff with sex.

Did he find it useful in his work to have been through that?

Nina Murray: I guess so. [She laughed.] I don't think so.

Do you really think Christiana committed suicide because of the breakup of their affair?

Nina Murray: According to their scenario she was to be the dying heroine, which also went with the kind of Wagnerian–Nietzschian folklore. They were terribly surprised and very upset when his first wife died before Christiana did. That wasn't according to the script.

How did Christiana's husband react to the affair?

Nina Murray: He died of tuberculosis, though basically, I think, of a broken heart.

Dr. Murray is pictured as the weak one of the couple and Christiana as trying to make him strong and tough.

Nina Murray: Nonsense. [She chuckled.] Everybody who knew them at that period has written me about it and said it's absolutely ridiculous—it's just the feminists' creed coming out.

Did he tell you of this relationship before you married him in 1969? [after the deaths of his first wife and of Christiana.]

Nina Murray: Oh, yes, all of it.

Next day I spoke with Henry Murray's biographer, Forrest Robinson, whose book Alfred Kazin described as an "astonishingly candid and detailed story of the bond between 'Harry' and Christiana . . . about the literary presences and psychological myths that dominated their lives."[3]

Why did you wait until Murray died before publishing his biography?

Robinson: Early in the interviewing process, which I started in the spring of 1970, it tended to inhibit people because they were fearful if they were candid, what they had to say would get back to Harry. Lewis Mumford said he would give me a very different kind of interview if he thought I was going to publish the book during Harry's lifetime. So I took that up with Harry and I said, "This is the problem," and we agreed it wouldn't be published in his lifetime. I had also read Robert Coles biography of Erik Erikson and thought it was very limited in value by the fact that he was clearly writing for Erikson alive, and there was an awful lot he couldn't do. I didn't want to spend time and effort when it was only going to be half true.

Was his idea for the thematic apperception test completely original?

Robinson: There were other projective tests around, so it was in the climate for those sort of things to be done. What was original was this particular type. A woman at the clinic had a part in it, Christiana had a part, and Harry had a part. It never gave him much pleasure to be closely associated with TAT. He didn't think it was a grand achievement, though it's still much in use and there are many different versions. He felt his grand contribution was the whole exploration of personality at Harvard in pre-World War II days.

Did Christiana have affairs with Mumford and Whitehead?

Robinson: She claimed she had affairs with everybody but had no empirical grounds for making those claims. I went over all the material very carefully and didn't find any basis for it. She has been portrayed as a liberated sexual animal and this may or may not be true. And I don't think her Nietzschian rites with Harry went on all that long. I think though that the one who really relished it and went for it was Christiana. Harry was much more conventionally heterosexual. Christiana had a much more complicated, passionate life than he did. I think she took the lead in most of that. She was sexually hungry and spurring him on to greater achievement in her letters. Some of it's real, some is rhetorical. There's a tendency in the Jungian frame of mind to inflate and to make things seem greater than they are.

Were they in love?

Robinson: I think so but their love involved this fantastic notion about themselves as the 20th-century incarnation of Dante and Beatrice— lovers for their age. They were going to be canonized for their love. They had a new way.

What was the reaction to the affair of Murray's first wife, Josephine?

Robinson: Horrified and mystified. She didn't have any easy way out of it. She certainly hated it but I don't think it was within the range of possibilities open to her culturally or personally to divorce. She was a conventional but decent, good person who didn't know what to do and I think she suffered terribly.

How about his daughter?

Robinson: Her life was very much influenced by Harry and wasn't too good. She's had a somewhat complicated life influenced by her father, and she never married. She's the image of her father. People would run into her in a train station in St. Louis and know who she was. She looks like him and sounds like him. She cooperated with me on the book.

How did Murray see himself like Melville?

Robinson: They were both deep divers in the unconscious.

Did Murray loathe Christiana at the end? [She drowned in shallow water in 1967 when she and Murray were together in the Caribbean.]

Robinson. There were awful difficulties in their relationship. She was a pathetic alcoholic and had to have this terrible operation for her high blood pressure in the 1940s which reduced her to a kind of invalidism. At the beginning of World War II, when Harry left for the war, they couldn't treat this life-threatening problem medically. So they had to use a terrible operation where they separate all the sympathetic nerves, and it's extremely painful. It was a terrible thing for a woman of her nervous type to have her energy reduced like that. So here we are some 25 years later [from the start of their relationship] and she's a full-blown alcoholic and very needful of Harry, although making herself unattractive to him in the way she's behaving. She's in a terrible bind and so is he, because he has this long history with her, their lives are all entwined. It's painful for him to watch this coming apart. It took something for him to stick it out. As Harry grew older he grew back into affection for his first wife. They came together again in many ways, so that meant that his relationship with Christiana was less intense than it once was.

Doesn't their relationship seem to you like folie a deux?

Robinson: Sure. You know, Jung is attractive to a certain kind of romantic sensibility. But neither Harry nor Christiana blamed him for what happened to them. They were both adults. They got what they came for.

Psychiatrist Anna Fels of the Payne Whitney Clinic of New York Hospital sees Murray as a "talented, wealthy man from a society family [who] pursued his fantasies on such a grand scale . . . while he maintained a conventional marriage and social life . . . that he wound up in a mythologized sadomasochistic relationship based on the Jungian idea of the symbiosis between animus, the male spirit, and anima, the female spirit. What emerges from this bizarre saga is a vision of Murray's disturbing intellectual and emotional limitations. For all his energy and enthusiasm, he left a swath of damaged lives: his wife's, his lover's, his lover's husband's, and more."[4]

Chapter Eighteen.
Cornelia Wilbur

"Multiple personalities are often brilliant, creative people."
—Cornelia Wilbur

The idea for the classic tale of multiple personality came to Robert Louis Stevenson in his sleep. Twelve years after he transmuted his nightmare into fiction, *Dr. Jekyll and Mr. Hyde*, Boston psychiatrist Morton Prince began treating a young woman with multiple personality. His book about this patient, Christine Beauchamp, was the first detailed and scholarly account of the successful treatment of the mental disease.[1] Therapists working today say that most "multiples" have endured overwhelming traumatic experiences, usually in childhood. By splitting into multiple personalities—often unknown to one another—they appear to be protecting themselves against complete and devastating mental breakdown or suicide.

Skeptics claimed that Prince and others who began to treat multiples had been tricked, that there was no such thing as multiple personality. Some said the patients were malingerers or mythomaniacs with an irresistible urge to lie. Others blamed therapists for inducing the multiple personalities under hypnosis when the patient was most open to suggestion.

Dr. Harold Pincus, a researcher at the American Psychiatric Association said: "Multiple Personality Disorder [MPD] is listed in the 1994 diagnostic and statistical manual for mental disorders, now renamed Dissociative Identity Disorder [DSM4] . The criteria are fairly explicit and there hasn't been a substantial change." They are:

1. At least two of these identities or personality states recurrently take control of the person's behavior.
2. Inability to recall important personal information that is too extensive to be explained by ordinary forgetfulness.
3. The disturbance is not due to the direct physiological effects of a substance such as alcohol.
4. It must be distinguished from malingering and fictitious disorder.

Psychiatrist Frank Putnam, a leading multiple-personality researcher at the National Institute of Mental Health, reports how one patient who reacted normally to a sedative drug in one personality, was totally unaffected by it in another. He also says that it is not uncommon for one personality to try to kill another, unaware that to harm the other is also to harm itself; while a third personality will prevent a potential suicide by calling the police.

Putnam recalled how in the mid-1970s, Cornelia Wilbur was speaking at a medical conference in Indiana about her work with multiple personality patients when, "a young psychologist began to heckle her. He suggested she had created this and that he could create it in his patients if he chose. Connie was a great pioneer. It was a very difficult time for her. There was very little appreciation of the effects of traumatic child abuse. Connie was one of the first people to understand the linkage between childhood maltreatment and the development of dissociative disorders like multiple personality disorder. So she was in a very hostile climate—also being a woman [she was at a disadvantage in a male-dominated profession]. But Connie was very strong and when the young psychologist challenged her she stood her ground and cut him down very smartly and won over her audience, at least for her pluck, whether or not they believed in the disorder. People who were at that meeting still refer to that years later—how she handled that heckler. I was vice-chairman of the task force on DSM this year and our report was published in May. It's considered a legitimate, recognized mental disorder that is well established. There are issues about it being appropriate in a particular individual, and there are groups of psychiatrists who do question whether it's a disorder primarily induced by medical care itself, through leading enquiry in highly suggestible people. But I don't believe that."

I asked: How about critics who say these patients are just good method actors?

Putnam: Nobody has ever demonstrated that's true. All the simulation studies that have been done suggest that there are real differences between simulators and actual patients.

By examination of brain waves?

Putnam: It's been done with a variety of different physiological measures, including memory studies, measures of electrical activity, and objective measures of visual acuity. All the simulation studies found significant differences between simulators and patients. In a recent review in *The British Journal of Psychiatry*, Augustus Piper, a longtime critic, acknowledged that from his reading of the literature that the condition surely existed but that it now becomes a question of is it being overdiagnosed? It's now more a question of how common is it?

How common is it?

Putnam: Probably the most systematic study was conducted in Winnipeg, Canada, in about 1991. They looked at 1,000 individuals chosen at random in Winnipeg and found one percent had pathological dissociation. That doesn't mean they all had multiple personality disorder. But they had some form of dissociative disorder.

Were almost all of them battered children?

Putnam: Dissociation is associated with trauma, and you can get significant levels of dissociation from many kinds of trauma. Recent studies have demonstrated high levels of dissociation in a subset of Vietnam veterans. Most studies of MPD find a very strong association with a history of childhood trauma.

Did any of the vets have MPD?

Putnam: Occasionally, but no one can say they weren't multiples before they went overseas. Like many people I didn't believe in MPD until I found it for myself. My experience is that until you find it for yourself you tend to discount it. She was a patient who had been hospitalized here for four years under another diagnosis and was discovered subsequently to have multiple personality disorder. That was a dramatic turning point, because she went from having no effective treatment for four years to being very rapidly discharged from the hospital. She's been doing very well now for the past ten years. So there was clinical proof that this was an effective way of conceptualizing this patient.

Had she suffered childhood traumas?

Putnam: Yes. She had a history of incest.

Was hypnosis the chief treatment?

Putnam: No, it was more psychotherapy and working with the individual personalities. She had been extremely suicidal and once we'd been able to identify which personalities were responsible for that behavior, it made it possible to work directly with them.

One of Cornelia Wilbur's former students, Dr. Francine Howland, a psychiatrist in New Haven, Connecticut, said: "A lot of us think the change of name from multiple personality to dissociative identity was merely political consensus. I think it's a real misnomer. If they'd said 'post-traumatic identity disorder,' I would have been much more pleased. In fact, we're hoping next time around that the committee will include something like this in renaming it. The work Cornelia Wilbur did on MPD described the disorder gorgeously and specifically. In some ways it's not a disorder at all, but a lifelong and very healthy response to overwhelming early trauma."

You say it's a healthy response. How about those who resort to violence?

Howland: I'm not sure that happens frequently. What happens is that as you become an adult with this particular response to early life trauma, your response is pathological. Whereas when you're a child and faced with overwhelming trauma, it's quite a healthy response. As you age the adaptive response in childhood becomes useless. It causes all sorts of problems in the real world. So it's adaptive in the sense that it works quite well when needed but it doesn't wear well in adult life. In that sense there are many parts of the brain, because of this, that are very disordered—cognitive and mood impairments, inability to track time—and that of course becomes problematic. Multiple personality was first described by Freud. Most psychiatrists would agree it exists. They might disagree about the numbers. It was once considered to be quite rare. I think we're coming to the conclusion it's not as rare as we thought. I have many patients who suffer from multiple personality disorder.

Are you a Freudian?

Howland: I was trained at Yale and have some psychoanalytical training. Connie [Wilbur] was a Freudian and worked from that vantage point.

How successful are you in relieving them of their distress?

Howland: If properly addressed it has an extremely good prognosis. These people have great brains and are extremely responsive to their environment. So when they're in a godawful environment, they adapt accordingly. So if you can help them change their environment they're responsive. That often becomes a lifelong endeavor. Some patients who have had less trauma you can integrate rather quickly. Others who have had overwhelming trauma perpetrated over years, take longer and can take a lifetime to deal with. People can become quite functional during their treatment, but they still struggle their whole lives with certain responses to old environmental triggers. It's just that they have a better grasp of things and are better able to cope once they understand where they're coming from.

How do you change their environment?

Howland: If you have a patient still living in an abusive household or relationship you get them to move away.

How did you meet Cornelia Wilbur?

Howland: I was a student of hers at the University of Kentucky from 1970 to 1974. Then I went to Yale for my residency. That's when I began to treat my first multiples.

So you've been at it for 20 years. I presume then that you agree with Dr. Wilbur that there is such a thing as a multiple personality disorder and that it can be effectively treated.

Howland: Absolutely.

Dr. Cornelia Wilbur is best known for her treatment of "Sybil," an account of which was made into a book and movie.[2] She told me of another patient named Jonah:

Wilbur: I have a beautiful film of Jonah, a black man who had four personalities. I love to show it because it astounds everybody. I am talking to the young man who has a gentle, quiet voice, obviously a very nice young man. Then I ask to speak to his alternate, a violent person. He opens his eyes, he hunches his shoulders, and he glares at me and says: "Whad you want?" And everybody gasps. It's so obviously different.

What made him a multiple?

Wilbur: He's had a terrible, traumatic childhood. His mother abandoned him to her drunken brother, who beat him regularly. And Jonah had a fanatically religious aunt who condemned him to hell. He was very bright. On the ordinary test, even with a third-grade education he had an IQ of 110. Multiple personalities are often brilliant, creative people. Billy Milligan, 140; a girl I treated was 185—140 is near genius. Jonah went to a country school where, like many black children, he ended up carrying the wood and water. He rode a school bus and three white boys, all bigger than he was, started beating him. They knocked him down and were kicking him in the ribs, and he was sure they were going to kill him. At which point he "vanished." Then this violent person took over, with superhuman strength, grabbed the smallest of the white boys and started beating the very skin off him. The other boys were so frightened by his violence and viciousness that they hauled their friend out from underneath him and ran away. So, whenever he got into a tough spot from then on the violent personality appeared and beat everybody up. He volunteered for the

Vietnam War, but Jonah never got to Vietnam. Instead, the violent personality got there and he fought the whole war and had a fantastic record. Finally, in a firefight he pinned down with one machinegun, the Vietcong, the Vietnamese, and the Americans; until one of his buddies persuaded him to let the Americans move to safety. So they took him out of combat and seven psychiatrists examined him and said he was suffering from battle fatigue. [She chuckled.] They sent him to Germany where he was given a landfill detail. That's a nice word for, "Handle the garbage, boy!" Which he was pretty annoyed by. But he went out and looked at a mountain of garbage. He figured he couldn't cover it with a bulldozer, so he decided to burn it down. He got a ten-gallon can of high-octane fuel and poured it all over the garbage, walked off the hill, threw the can on top and threw a packet of lit matches after it. And, because the thing blew up, they had garbage all over West Germany. They took him off landfill detail and he saw six more psychiatrists. [She chuckled.] None of them made the diagnosis of multiple personality. But they sent him back to the States [laughed] So after being seen by 13 psychiatrists he was finally discharged with honor and a bunch of medals. He came to Lexington and got into a fight with his girl and got arrested. The police thought he was talking funny and brought him to the hospital. And he was talking about himself in the third person. One of the residents said: "I think I've got a male multiple personality, a dual, anyway." I said, "All right. Admit him." So he did, and [eventually] I said to the resident, "He turned out to have four personalities."

Are the EEGs sometimes quite distinct for each separate personality?

Wilbur: We sent Jonah up to the EEG lab to have several EEGs done and our neurologist said: "Well, I'm happy to do it, but I think it's pretty silly. Because it's the same brain and you'll get the same brainwaves." He came down two days later and said, "I'm a convert and I feel kind of sheepish, because I found five different EEGs." I said, "Five! the man only has four personalities!" And he said, "You must remember, when people are awake their EEG is different from when they're asleep. When this man's awake he has one personality, and in a hypnotic trance he has another one. But his three other personalities, produced spontaneously or by trance, have different EEGs from either the waking or the sleeping person."

How did he respond to treatment?

Wilbur: The chairman of the department at the University of Kentucky asked me if he could have Jonah as a patient and I said he could, but he never treated him to recovery. But he was very much better: he understood his condition to some degree.

What happened to him?

Wilbur: He was killed in an auto accident. A drunken driver collided with a car in which Jonah was a passenger.

When the therapy is effective and the personality is fused, and then faces a traumatic experience similar to the one that traumatized him or her into a multiple personality, is there any danger of disintegrating again?

Wilbur: Near to the point of fusion, yes. They may come in and say, "Well, I'm all one this morning," and then something may happen and one may split off. But they know that. They'll say, "Rita took off because . . . but she'll be back." And she will. Because the entire complex is stronger when they are fused than any one of them is when they are not fused.

Are they completely safe then?

Wilbur: Well they may be fused and then may defuse if there are serious traumatic situations. That usually only occurs shortly after the first or second fusion. They may come in and say, "Well, I'm solved. I know I'm not going to separate again." Or, "I'm never going to have to use that as a defense in the future." You can trust those statements. They know then that they're not going to come apart. Multiple personality is a very effective defense. Just great. Sybil couldn't get angry at her mother, because you had to honor your mother. So Peggy Lou [one of her personalities] got mad as the dickens and stomped around saying, "Mrs. Dawson is not my mother!"

Your aim is always to fuse or integrate the multiple personalities?

Wilbur: Right.

It took you 11 years to cure Sybil. Was that the longest treatment to a cure?

Wilbur: Yes. But when I started out I didn't know she was a multiple. I thought she was a hysteric. When I found out she was a multiple, I went to the available literature. What was available then? Just Morton Prince's thing about Miss Beauchamp and his experiments with her and *The Three Faces of Eve*.

[Eve—Mrs. Chris Sizemore—believed the trauma that first triggered her multiple personalities was seeing her mother cut her arm. Fearing her mother was going to die, instead of running to fetch her father, Eve buried her head under a pillow. Then she seemed to recede in space as she imagined she saw another little girl go to get help. Other traumatic incidents were when she saw two corpses: a man who had drowned in a ditch, and another in a lumber mill who had been cut in half with a saw.]

I know that Eve, who eventually had 21 personalities, would stop in midsentence as one personality, then there was a moment when she felt there was no personality at all, and then she would complete the sentence as another personality. Did Sybil also change before your eyes, and without warning?

Wilbur: Sometimes. And it got so that I could predict when she would change, excepting under one circumstance. If she was thinking about something and had not yet started to talk about it, and it was very disturbing to her, she might change from one personality to another in an attempt to find the personality who would be able to tell me about it more comfortably.

And that personality was cooperating with you, the therapist?

Wilbur: That's right. And Vickie was the most cooperative personality. Sybil used to get so cross because she'd be afraid or embarrassed to tell me about something, or have other defenses, or resistant feelings. Then I would say quietly, "Vickie." And Vickie would come. I'd say, "Can you tell me about it?" Vickie would say, "Yes, but you'd better be careful how you use it when you let Sybil know about it." Then Vickie would tell me what was troubling Sybil.

In a trance?

Wilbur: Spontaneously. Later I'd mention to Sybil what Vickie told me, and Sybil would say, "Where did you get that information?" I'd say, "Vickie told me." And Sybil would say, "That blabbermouth!" [She laughed.]

Were her dreams different for different personalities?

Wilbur: Yes. Peggy's were always off on some kind of Sherlock Holmes pursuit, finding and locating things. Sybil would have concentration-camp dreams one after another. She identified with Hitler's victims, because her home was like a concentration camp. And she would turn up in her dreams with numbers tattooed on her arms, all kinds of dreadful things. None of the other personalities had concentration-camp dreams.

[Sybil's mother was a schizophrenic who tortured Sybil as a child, throttling her with scarves, burning her with irons, locking her in a trunk, mutilating her internally with button hooks and other objects. Her father did nothing to protect her, never asking, for example, why she had a black eye or fractured larynx. Her parents regarded any pleasure as a sin, though they exposed her to their own sexual intimacies by having her sleep in a crib in their bedroom until she was nine.]

Were her other personalities more carefree than Sybil?

Wilbur: No. Sybil would say, "What do I have to live for?" But Marsha would get actively suicidal. It was kind of scary a couple of times. One time Vickie called me and said Marsha was going to commit suicide. I said, "Can you keep her there until I get there?" She said, "Yes, I think I can." And she did. Now Sybil is able to recall all her experiences—every last one—and be fairly detached about it. I sometimes think she is a mentally healthier person than the so-called average person, because she's been there and back. Now nothing frightens her.

Is it ever a problem to discover which is the basic, core personality?

Wilbur: I don't think so. We had been working with a young woman who had three personalities, and when she came in she said she was Barbara. After she was here a few days, Barbara completely disappeared, and we didn't see her for three weeks. She had another personality who talked very freely, and I said to her, "Who's the original?" She said, "Oh, Barbara is. She'll come back." [She chuckled.]

As a Freudian I suppose you accept that for most normal people the knowledge of their unconscious is kept from them by their normal minds.

Wilbur: Yes.

Can multiple personalities always be cured?

Wilbur: It's my own experience that they're very definitely curable. Freud had a considerable influence on me in a funny way. Granted he treated a lot of people and I'm not sure if he cured any of them, but I took his word for it that basically people hide things from themselves. These things sit there in the unaware parts of their minds and cause trouble. They say that 98 percent of motivation comes from underneath. I'd say that the unconscious is awfully powerful and what we do is to accommodate it. I once had to see two women patients, two hysterics. One could only whisper and the other was a very attractive young bride who was paralyzed from the waist down. I listened to them and they told me exactly what was the matter. And I said, "Look, you have a conflict, and this is what it is." The girl who couldn't talk—that was her problem—said, "Is that so?" in a loud voice. The girl who couldn't walk, got out of bed and paced around and said, "Well, that makes me feel bad." So then I was convinced. I can't guarantee that I did exactly what Freud did, but I got the idea from him.

Have you simplified your account of how you cured them or did it happen exactly as you described it?

Wilbur: One happened almost exactly like that. The other took a little longer. And I [she chuckled] developed a terrific reputation for being a magician with hysterics. So I have always been convinced that hysterics are perfectly curable. I took an hysterical man in Lexington and in a week had him understand why he was paralyzed in both legs. I sent him home after only eight days of hospitalization.

How did you know the multiple personalities were not simply putting on an act?

Wilbur: Every now and then I get a dual or multiple personality on the way to getting well and all of a sudden they turn on you and say,

"I've been playing a game." I say, "Why are you saying this at this point?" "Well, I'm tired of the game. I just wanted you to know that I was doing this to fool everyone." So [She chuckled.] when you get beneath this, it's one of their final defenses against revealing things that really bother them. Anyone who falls for this sort of thing is a fool. As a matter of fact, Sybil fooled us, too. She wrote me a letter and said she'd never had any amnesia, etc., etc., etc. I knew this was an outrageous attempt to get out of going ahead.

Is the handwriting different for each personality?

Wilbur: Oh, yes. It's very interesting. Flora Rheta Schreiber, the author who wrote the book about Sybil after getting the material from me, is a linguist and speech teacher among other things.[3] When she was given the tapes of Sybil talking, she let out a small yell and said, "It's different people!" She got so she could tell who was on the tape simply by their voice characteristics. Now she says that linguistic characteristics are completely different, the diction is different, and the pronunciation is different. She points out, for example, that Peggy Lou spoke and pronounced words like a country girl. She'd say, "You *jist* don't give me a chance." And that Vickie always used French phrases, and Marsha and Vanessa both talked with a British accent. The family on both sides is originally English. The personalities are different in every possible linguistic way. The same was true of body movements. I could tell who they were, after I really got acquainted with them, by the way they walked in the door. Marsha and Vanessa were like a pair of twins who, though alike, were distinguishable. The others were so totally different there was no problem. They liked different foods and they ate different quantities. Sybil ate very sparingly. Peggy gobbled her food and ate a great deal. Vickie would eat very well, but in a ladylike manner, and always leave a little on her plate because that was proper. [She laughed.]

Can you explain your success with multiple personalities?

Wilbur: I think my ability comes from a great deal of knowledge that is more available to me than to most people. There's nothing mysterious about it, partly through my training. But I have also learned to trust my feelings and my unconscious.

Do multiple personalities manifest the same behavior in a hypnotic trance as spontaneously?

Wilbur: Yes, exactly.

Have you cured any since Sybil.

Wilbur: Yes, I've fused several. One of them in about three years.

Are you working with any now?

Wilbur: I certainly am: five. Two of them are violent and both as children thought they were going to be killed.

Is it very exhausting work?

Wilbur: Yes.

Have you made any mistakes in treatment?

Wilbur: Yes. Back in the beginning I came close to addicting one patient to pentathol. And when I was treating Sybil I didn't really know what I was doing, I didn't know specifically what direction to take. Having all of that experience, I am much more efficient.

Would you cure a Sybil today in less than 11 years?

Wilbur: Oh yes. In five years or less.

What is your advice to psychiatrists treating multiple personalities?

Wilbur: The more continuing medical education they can get, the better. There are all kinds of courses now in relation to this. There are advanced courses given by the American Psychiatric Association. And I'd tell them to diagnose these patients in two or three years instead of the average six-and-a-quarter years it takes now.

But if EEGs demonstrate multiple personality, isn't that a sure, swift and simple diagnostic method?

Wilbur: No. If you're going to get an EEG on an alter [alternate personality], you've got to find an alter. And if one person comes in to you and you just sit and talk with him and he doesn't change, then you haven't the ghost of a chance.

How can psychiatrists get alters to appear?

Wilbur: By getting the patient's confidence. You can do a good psychiatric interview and it will give you the proper diagnosis. You can also use hypnosis.

Are you considered the leading expert on multiples?

Wilbur: Yes, but there are others who are certainly as expert as I am.

How many multiple personalities are there in America?

Wilbur: I'd guess there are a minimum of 5,000 in the nation.

Psychiatrist Frank Putnam of the National Institute of Mental Health admires Cornelia Wilbur as a pioneer in the diagnosis and treatment of those with multiple personality disorder. When I discussed the subject with him in November, 1994, he stressed that in the United States all psychiatric diagnostic categories are established by professional consensus and the MPD meets all the same requirements demanded of other psychiatric disorders. In his research Putnam found that some MPD patients had as many as 100 alters and that the different personalities even differed in allergies, right-or left-handedness, and the need for eyeglass prescriptions.

Chapter Nineteen.
Ashley Montagu

---

*"As an anthropologist I know that a great many beliefs of many
so-called primitive people—they're not: we now call them indigenous
people—are vastly more intelligent than most of the beliefs of the
civilized peoples of the world."*
—Ashley Montagu

Anthropologist and social biologist Ashley Montagu takes pride
in having changed Albert Einstein's mind about a long-held belief,
and he told about it in our conversation. Montagu has expertise in
genetics, paleontology, biology, sociology, and anthropology. In fact,
ask him almost anything about human nature and he has a persuasive, well-informed answer. Born in England in 1905, from which he
emigrated in 1927, and educated in England, Italy, and the U.S., where
he obtained a Ph.D. from Columbia, Montagu has taught at Harvard,
Princeton, and at New York University. He has written over 40 books,
including *Man's Most Dangerous Myth: The Fallacy of Race, The Natural
Superiority of Women, Life Before Birth, The Science of Man, Man Observed,
Immortality, On Being Intelligent, The American Way of Life, The Anatomy
of Swearing, The Dolphin in History* [with John Lilly], and *The Prevalence
of Nonsense* [with Edward Darling]. Now revising three books and
writing a fourth on shock, Montagu said he worked more than my

24-hour day. "You know how you work 25 hours a day?" he asked. "You get up an hour earlier."

Tell me about your meetings with Einstein.

Montagu: I was shy about meeting him. I have been diffident all my life about intruding on great men. I regret that very much now, because I missed many opportunities where I know it would have been not regarded as intruding by the likes of Havelock Ellis and H.G. Wells, and such characters, to whom I had access. But I never took advantage of it, feeling I was a mere worm in comparison with their enormous stature. This is why I never wrote down anything Einstein said to me on the occasions when we met. I felt these were simply meetings to talk about mutual interests.

You first phoned him and he invited you over.

Montagu: Yes, I was planning a film on the dangers and possible benefits of atomic energy for the American Federation of Scientific Workers, and I contacted Einstein for advice in June 1946. I very soon realized that any idea, however contradictory to any he may have held, that was sufficiently explained and given scientific underpinning would be one that he would accept. And this was true of Einstein's belief in the innate depravity of man, "original sin," as it were; that man is born an aggressive creature. There is a famous letter that Einstein was induced to write to Freud, "Why War?" And Freud answered that there was an instinct in human beings towards destructiveness and war. And he was very much afraid that there was no cure. He went on to say, "I have heard that there exist people in the world who know no conflict, but I've grave doubts about that." So Einstein accepted the general view that man is indeed endowed with instincts. When I first explained to Einstein that human beings have no instincts, he thought that was really quite outrageous. But eventually he accepted the idea. I simply defined for him what instinct was. He came up with things like the sex instinct. And I said, "That's the easiest one to dispose of."

But surely the sex drive is an instinct. Isn't it necessary for our survival?

Montagu: It isn't necessary for survival. Not everyone has to indulge in sex for a species to survive.

Here's a definition of instinct in *Webster's New World Dictionary*—see if you agree: "An inborn tendency to behave in a way characteristic of a species. A natural, unacquired mode of response to stimuli; as suckling is an instinct in mammals." Isn't suckling an instinct in human babies?

Montagu: No, it's certainly not an instinct; and let me say this, and I hope it will contribute to the enlargement of your mind as well as the increase in your knowledge. A dictionary is not a good source for understanding the meanings of a word, for the reason that it's likely to be apostolic. In other words, to have recorded the wisdom or unwisdom of the period in which it's compiled. So you'll find in practically every dictionary a definition of the sort you've just quoted. The fact is, for example, insofar as suckling is concerned, babies are not instinctively capable of suckling. Nor are chimpanzees, nor gorillas, nor orangutans. They all have to learn to suckle. And suckling should not be identified with sucking, which is a very different thing. If babies attempted to take their nourishment by sucking, the species would have been extinct long before this, because they wouldn't have got much. Suckling is a very interesting phenomenon for the reason that it's been around ever since the human species has been around. Yet no one has ever paid any attention to what a baby actually does at its' mother's breast, except myself.

I have published on this, but I sometimes feel that I'm like a giant dynamo discharging into a nonconductor so far as the effects are concerned. In brief, what the baby does is to take in the whole of the areola region—the nipple is not what he sucks—that lies free between his tongue and his hard palate, and he turns his mouth into a suction pump, a hermetically sealed suction pump. That's the way he suckles. But he has to learn to do this. Now, it may take him a few minutes, a few hours, or a few days, but he has to learn to do this: learning being defined as the increase in the strength of any act through repetition. As for sex, everyone has to learn. Monkeys, apes and humans all have to learn how to conduct themselves in the matter of sex.

How about the first humans or monkeys? Who taught them?

Montagu: I don't know that anyone taught them, except they observed other animals and went and did likewise.

So that if it isn't immediately done, even if it takes only seconds or minutes to achieve, it is not an instinct. Is that it?

Montagu: Very definitely. And I pointed this out to Einstein. Then, on the matter of war and innate aggression, the belief that one is born with an aggressive instinct in human nature was congenial to him. He started out with that idea, but finally was convinced he was wrong. This, of course, pleased him, being such a proponent of peace.

I told him that an instinct is very simply defined as an innate psychophysical disposition to receive a particular stimulus in a particular manner, and to react to it with a particular behavior accompanied by a particular emotion. That's an instinct. And human beings have no instincts. And when Einstein asked, "What about the sex instinct?" I said, "Sex is a complicated drive which develops gradually and which is in no way an instinct. It is something that has to be learned."

Two babies, a boy and a girl, shipwrecked on an island and who grew up together, would either commit suicide, or they might by pure experiment discover how to do it. We have all the anatomical and physiological equipment, but we don't know what to do with it unless we learn. It's like playing the piano, or dancing.

When Einstein said he believed man was innately aggressive hadn't he heard of peaceful communities?

Montagu: I wasn't surprised that he knew nothing about the peaceful Hopi or Zuni Indians, or the pygmies, because there were only certain areas in which he read. He obviously didn't know much about international law, either. Because when I asked him how we could get people interested in seeing that nuclear energy wasn't misused, he suggested international law. I said, "Professor Einstein, international law exists only in the textbooks of international law." He called that a really outrageous remark, then thought for at least two minutes, and said, "You're quite right." I had to tell him that of all the treaties that have ever been signed between nations, every one has been broken with the exception of the one about the borders between Canada and the United States.

He didn't read much outside the literature of physics because that's a full-time job if you have the kind of mind Einstein had. In other words, he was thinking all the time about a great many matters which had been unresolved, and which were in the realm of theory and experimentally unproven. So he was occupied. And I understand it. One goes about in "a brown study" used to be the phrase, and doesn't have much time for extraneous reading.

This is an account by writer Max Eastman of first meeting Einstein: "I usually receive from great men in any walk of life, and always with surprise, an impression of feminity. In Einstein this was striking, his gestures being fastidious rather than compelling, and his swaying gait suggesting to my mind a buxom mammy. His hands were fat, veinless, and unwrinkled like those of a baby. It was a strange body in which to locate a mind with an edge so keen and hard it could penetrate all the natural assumptions of life and conversation, and even of what had been physical and chemical science, to an armature of mathematical abstractions which contain hardly a recollection of the concrete perceptions that make them valid."[1]

Montagu: That's exactly the impression I received of Einstein. The way I'd describe him: he was like a good Jewish mother. Bodily he had a softness and roundness about him—not intellectually or cognitively, of course. And he walked as if on air. All my life I've been very much interested in dancing. In fact, it's been said that now that Fred Astaire has gone I've replaced him. It was very striking. There's a long corridor in his house and he was at the other end when Miss Dukas called him and announced I was there. He cordially invited me up to his study. And then he seemed to glide toward me in a sort of undeliberate dance. It was enchanting. It was, maybe, the way someone else might whistle as they moved. He danced. He was very fond of music and the violin, and he seemed somehow to be expressing his love of music as he walked.

Someone said he had a woman's voice.

Montagu: That I did not notice. It certainly wasn't loud, and it was a very pleasant voice. One thing I noticed about him was that on the telephone his voice was always unusually clear, as certainly was his speech. I had read a lot of theoretical physics in the days when relativity burst upon England in the 1920s and I read about it a good deal later, too. So I could talk to him about a subject of interest to us both.

In your book *Immortality*, you wrote, "Since men's thoughts are a reflection of their souls, of themselves, their souls, in this sense, may
be said to live on in their thoughts, in their deeds, in their works. And
this, I believe, is the only real immortality."[2] By "souls," of course,
you don't mean in any religious sense.

Montagu: No, not at all.

You mean the character and personality.

Montagu: Yes.

Did you discuss this with Einstein?

Montagu: I don't think so, but he thought of it in much the same way
as I do.

How did he behave during the McCarthy period when he was accused of being a Communist?

Montagu: I called him up at once and said, "What do you think you
ought to do about this? You ought to sue the slanderer." He laughed
and said, "Good gracious, no," and words to the effect that such things
were beneath his notice. I was accused of being a Red, too, because I
was a member of a Committee of the Arts, Sciences and Professions
to Re-elect President Roosevelt. That's in my FBI files. Neither of us
had to face the un-American Activities business, but if it had gone on
long enough we probably would have had to. I had my passport taken
away from me by the State Department. I'd just returned from a meeting of UNESCO, where I'd been helping draft a statement for the
United Nations, and I received a phone call from a man who said he
was from the FBI and could he see me. And he did, and said, "I've
been instructed to lift your passport." I asked why, and he said he
didn't know. So, like a fool, I gave him my passport, which I really
shouldn't have done, and didn't have to.

Then I wrote a letter to Mrs. Knight, who was the ogre who
presided over the proceedings at the passport office, and asked to explain the behavior. She wrote back saying it was because I was, among
other things, a member of the Committee to Re-elect Roosevelt. I
wrote back, "If you don't return my passport by next mail I shall make
a public issue of it." So my passport came back in the next mail. Bullies

are, as I know from experience, people one should stand up to immediately. It was during this period that Einstein spoke and wrote most pessimistically about the United States, shortly before his death.

Do you think the expectations from examination of Einstein's brain are realistic? That studying his brain cells will give clues to his genius has any validity?

Montagu: None whatsoever. I'm an anatomist by training, and the brain was one of my principal interests. You can't tell the difference between the brain of a chimpanzee and a human being from its form or its microscopic structure. You certainly can't tell anything from the microscopic structure of the brain whether the person was an idiot or a genius. Lincoln said, "God must have loved the common man; he made so many of them." He didn't add the corollary that God must have disliked the uncommon man, because he made so few of them.

Preserving Einstein's brain to study was Dr. Thomas Harvey's idea.

Montagu: He kept the brain in a jar in the basement of his house. Then he separated and got divorced, and the brain was still down there. That's how I learned about it. I was shocked to discover that there was such an organ hidden in a jar down in [the former] Mrs. Harvey's basement. She said, "I wish someone would get rid of the damn thing." Eventually Harvey claimed it.

[Dr. Harvey disputed this account, with a laugh, saying: "I don't think it was ever kept in the basement of my house in Princeton. I had it at Princeton Hospital, and when I left there I did take it with me, of course." When I told him Montagu's opinion that trying to find an explanation for Einstein's genius by examining his brain was a futile enterprise, Harvey replied, "Some scientists believe you can learn about the function of the brain from its structure. (Roger Sperry didn't.) Dr. Marian Diamond, a professor of neuroanatomy at the University of California, Berkeley, (who examined part of Einstein's brain) says her research indicated there was a difference between Einstein's and the average person's brain. Whether it shows more intelligence or not, remains to be seen."]

Dr. Harvey believes it may be possible to learn about the function of the brain from its structure.

Montagu: It's a tribute to the ignorance of medical education and medical men, because anyone who knows anything about the subject knows that you can't tell a thing from the structure of the brain about the quality of the mind. Removing it was due to ignorance as all stupidities are, although Einstein might not have minded. The normal brain is about 1,350 cubic centimeters, or three-and-a-half pounds by weight. The largest brain on record is 2,500 cubic centimeters—almost twice the size of the normal brain. That was the brain of a hydrocephalic, that of an idiot. Even a gross feature like the size has no relation whatsoever to intelligence. Even the most sophisticated research can't make discoveries about the normal brain through examinations.

Dr. Marian Diamond reported that Einstein's brain had more glial cells for every neuron than do average brains, in a region of the brain known to be involved in the analysis of information. Does that make sense?

Montagu: None at all. I have great respect for Marian Diamond, but she's wrong.

I spoke with Montagu again a few years later on February 24, 1995.

What are you working on?

Montagu: Several things at the same time: the sixth edition, fortunately, of my book, *Man's Most Dangerous Myth: The Fallacy of Race.* [Montagu had drafted a statement on race for UNESCO.] And I'm also trying to work on a book called *Stress and Shock.*

I have interviewed Hans Selye, the expert on stress. Will you quote him in your book?

Montagu: Most certainly. I knew him very well in two phases of his life: one, when he was insufferable, and the other when he became a very lovable human being.

Maybe he changed after they gave him artificial hip bones.

Montagu: That's right. It changed his life.

Francis Crick's conclusion, after studying the brain, is that consciousness, the so-called spirit or inner man, and all thought, is merely the firing of neurons in the brain. The mind is purely physical.

Montagu: That declares him at once for being ... captivated or trapped by the imprisonment of intellectualism.

He's too sweeping?

Montagu: Of course.

Because it's largely unknown territory?

Montagu: To a very large extent. But one thing is clear. It is that brain through which the passage of an infinite number of possible experiences can pass which makes a human being. And the culture results from that. He is conditioned in it and becomes either a bigot or an open mind, but not so open that his brains fall out altogether. The fact is that the more important part by far of a human being is his humanity, to relate not only to all living creatures, but being—what used to be called by people who were regarded as eccentrics—in tune with the infinite. My last book will be with my physicist friends who will be providing the physical basis for this statement—a field theory. The title is *Three Men in a Balmy Boat*, something like that. Only it will be two men writing the book.

Do you then say it's not possible to justify everything about human nature by physical facts?

Montagu: Oh, yes. Physical facts are not good enough. Human beings are the only creatures who speak, have a language. And language, as you probably know, is the beginning of falsification. In other words, this is where we can go way off the beam and turn it into a system of reality to which we all subscribe, and it's provable within the terms of our understanding of what reality is, without being aware that all reality for human beings is socially created. By that I don't mean the universe is socially created, though part of it is understood in such terms. I mean that everything we do and know as human beings is socially created.

Most scientists seem to be atheists or agnostics. What are you?

Montagu: What I personally think is this. If England had a climate I would have been living there years ago. And I would have been living near a Saxon church, one of those square-tower churches on which the roof has fallen in or collapsed, as almost all of them have. And living in a thatched-roofed house near a family in which there was a good English cook. The best cooking I've ever had in the world has been domestic cooking in a few English working-class homes. There is nothing like it anywhere else in the world. I would go to church, certainly on Sundays. And I regard myself as a born-again atheist. But when I go to England, that's where I go.

As a born-again atheist why are you going to church?

Montagu: Because it has a spiritual quality within it that . . . It's difficult to put into words.

You go for aesthetic rather than religious reasons.

Montagu: No. They're religious reasons. I do feel in tune with, let us say, something.

What do you think of Murray Gell-Mann's view that everything in the development of the universe since the Big Bang has happened by chance or accident?

Montagu: That's where he's wrong. He's wrong as a physicist. It's just not true. If what he wants to say is that evolution is the maximization of the improbable, I think it's fairly good. And, furthermore, in physics we can show there is a field theory possible which shows that everything is connected with everything else.

Gell-Mann would probably agree with that. [Nobel physicist Murray Gell-Mann also says that *everything* is made of the same fundamental particles—quarks.]

Montagu: And furthermore if you want to call that God, you may. What Gell-Mann calls an accident, I think must be seen as design. Not that there was a designer. You probably know Dawson's book on the blind watchmaker—a comment on Paley's famous book*—if you were walking about and found a watch you would have to conclude there was a watchmaker. But there is something that develops which is a design, which you could get almost any physicist to agree to. I mean it doesn't hold together if there isn't. But it's not an accident. And the Big Bang is not an accident, either.

* William Paley (1743–1895), who taught at Cambridge. The watchmaker is mentioned in his book *Natural Theology*, Frederick Ferre, ed., Indianapolis, 1964.

Let me get this straight. Am I quoting you correctly as saying, "You can call that God"?

Montagu: That makes me sort of a victim of the the religious right: here's Ashley Montagu affirming the existence of God. And that's not what I mean. There's a book called *The Anthropic Principle* in which this idea is put forth that there is a design.

Does the design also mean a purpose?

Montagu: No.

So when you says "You can call it God," you don't mean a personal God. What's puzzling, though, is this: if the creation of the universe wasn't an accident—which a layman like me thinks means without intention or purpose—then how is there a design?

Montagu: Design can come about as in the case of snowflakes. Look how beautiful they frequently are. They look as good as and often better than any human designer could manage. But there is no other force behind it than certain physical laws with which we're very well acquainted. That is my opinion: there's no designer. Take the structure of the atom, for example. Here you find a number of elements which were postulated by mathematicians, not by physicists, who postulated that these elements must exist. And you can call them whatever name you like, neutrino or whatnot. We're continually inventing new names to explain these things, but the fact seems to be that what many people describe as an accident isn't really an accident at all in the sense that they understand it—that this was something unpredictable. The fact is that most people at the end of the last century in physics were willing to jettison the idea of the atom, because they felt this was merely a series of accidents that had occurred, and that matter would come into existence in its various forms by pure accident. And by pure accident, you mean generally that you can't explain it.

When you say, "You may call it God," you mean, "I, Ashley Montagu, call it nature or nature's laws."?

Montagu: Yes.

How about those who believe the universe was made out of nothing?

Montagu: The joker in the pack there is what they don't take into consideration, that only nothing can come out of nothing.

As Shakespeare said.

Montagu [chuckled]: Yes.

So are they being ridiculous in calling it "nothing"?

Montagu: Of course. Furthermore, our thinking about these matters should be at best problem solving, which all thinking should be. And we do our best to think this way through mathematics and physics. However our language limits us to certain ways of putting things, of understanding things and we become prisoners of our vocabulary, which asserts that when I put a piece of wood on the fire it is burnt to ashes and that's an end to it. Well, there isn't an end to that either, because of course in the process you've introduced a large number of changes in the world which are invisible to your naked eye. But you've produced a lot of changes which will have environmental effects and go on ad infinitum.

You don't believe in life after death.

Montagu: No. I've seen enough of what happens to human beings physically and know there are all the claims about your soul—when you die you weigh a little less, and your beard continues to grow. All that is explicable very easily on purely physical grounds. Most people believe in immortality when they're young. They think they're going to live forever.

Have you read any books on near-death experiences?

Montagu: No, because I know what they are, too.

How about the brilliant white light so many experience?

Montagu: I had this experience. I've written about it. It was the effect of an anesthetic. I not only had that sort of experience but I actually met God and had a conversation with him under circumstances which could not have been disputed. It was a time when I was very concerned about the way things were going in the world, during the Spanish Civil War. I, of course, was on the side of the people who had elected their own government, and Franco was trying to destroy it with the aid of the

Americans, Germans, English, and French. [Their arms embargo put the Republican government at a disadvantage.] I was worried about the state of the world. And God, who looked like Father Christmas and was a very gentle soul called me to his side and said that if I would stand by in the shadow I would observe a visit from all the leaders of the world. And, of course, they came in one by one. You had a simple question: if you had a choice of doing the right or the wrong thing, what would you do? And they all answered, of course, "the right thing." And when the anesthetic wore off I was so happy that I told my anesthetist, "This has been a wonderful experience." And he said, "You'll be all right." And I said, "No, this is the real thing that happened. A marvel. Reality." So I know that under certain conditions you can experience a reality which is even more real than the unreal, as that was to me. And people who have after-life experiences are in that state where they also, the same as UFO captives, really believe it happened. But it's all a dream.

A Harvard psychiatrist, John Mack, has written a book about UFO captives.[3] Some of these books are written by medical doctors who deny anesthetics or drugs or the imagination are contributing factors, and say the experiences can't be explained away.

Montagu: I haven't read any of these books. Like a good bigot I just feel I know what the answers really are.

A lot of them feel consolation and have no more fear of death.

Montagu: Well that's not bad. They are going to die, nevertheless, in the conventional sense. I know as an anthropologist that whole peoples believe in reincarnation as firmly as they believe in anything. In fact, if it weren't for their belief in reincarnation the settlement of Australia could not have been as easy as it was. When the aborigines first beheld white men they welcomed them as long lost returned members of the tribe, because they indulged in tree burial and if you're exposed in trees for several weeks or months your whole body loses its pigment and you turn white. They are highly intelligent, sensitive people and they believe this. I really ought to write a book about it, but I haven't got enough time. I'm 89 and I'm only going to live to 110 years, according to present diagnostics.

But the fact that so much is mysterious, consciousness, say. Where was our consciousness before we were conceived? And the fact that maybe there are many universes unknown to us. Distant galaxies. The mere fact of birth is so extraordinary. The surviving of bodily death doesn't seem all that extraordinary, in those terms.

Montagu: Quite right. Again, as an anthropologist I know that a great many beliefs of many so-called primitive people—they're not: we now call them indigenous people—are vastly more intelligent than most of the beliefs of the civilized peoples of the world. They have a coherent system of beliefs which make absolute sense. And you cannot change their ideas because they make such good sense.

Do they still exist?

Montagu: There are a few Australian aborigines left, and one or two groups of Eskimos in the Aleutians. I've written a book called *Immortality, Religion and Morals* in which I discuss some of these ideas. Sir James Fraser has three volumes on the belief in immortality which are fascinating, like everything he wrote, and haven't been read by any anthropologist for 70 years or more. Pretty wonderful. And again, the important thing for a writer, as you probably know, is not so much that he is right or wrong but whether he is fruitful. You can be right and not very fruitful, and wrong and very fruitful.

Who's wrong and fruitful?

Montagu: Well Sir James Fraser is believed by many anthropologists to be wrong headed. *The Golden Bough* is described by some anthropologists who haven't read it as like Kipling's *Just So Stories* [fiction]. They are stories which are firmly believed in and which are a very real and important reason for our longevity. I've known people who told me that if they didn't believe in God they would find it impossible to go on living. Belief in immortality or reincarnation has served a very useful function in that among other things it enables people to go on enduring what they have had to endure. And the belief that in the afterlife they're going to have a much happier existence. So you can distort this as the Christian fathers have, into a sort of emotional challenge to human beings who must believe, because if they don't they won't go on to another life. It's very conditional this reincarnation and immortality upon your following the teachings of the particular religion; in this case Christianity.

By "emotional challenge" of the Christian fathers you mean they know it isn't true?

Montagu: Yes.

You could explain most martyrs that way, too.

Montagu: Of course, and most heroes in battle. I lived through the first World War and saw Victoria Crosses being pinned on the tunics of these heroes. Of course, what some of these heroes really believed was they were actually dying for their country when they ran forward and faced several machine-guns at the same time and by sheer luck managed to survive. They became great heroes and were convinced that they were going to be rewarded in the other world. But began to doubt it very much when they found they were unemployed. I remember the slogan, "England—the Land Fit for Heroes to Live In." They ended up standing in the gutter selling apples, or lying on the pavement with chalk and attempting to draw various figures, with the cap for money to be thrown into. And this goes on. The brainwashing of millions all over the world who become the willing cannnonfodder, dying for their country, or dying for people like Mr. Nixon or Mr. Bush.

I suppose you could say people can live without a belief in God or the hope of immortality from the way many Soviet Russians lived presumably reasonably decent and productive lives.

Montagu: Of course there are many kinds of Russians and all through the Communist years there were Orthodox Christians who really believed but kept quiet.

Will your next book suggest a more ideal way of living?

Montagu: All my books do. I'm accused of trying to save humanity from its debacle, and this is why for about 45 or more years I've been interested in babies, infants, and children.

Chapter Twenty.
Wilder Penfield

*"One can also recall the poignant questions by computer lovers:
At what stage of complexity and performance can we agree to endow
them with consciousness? Mercifully this emotionally charged question
need not be answered. You can do what you like to computers without
qualms of being cruel!"*
—Sir John Eccles, Nobel prizewinning neurobiologist.

As Freud and Jung are still regarded as masters in their field, so Dr. Wilder Penfield, almost 20 years since his death in 1976, is still cited by experts as *the* authority on the brain who speculated about the possibly transcendental powers of the mind.

Penfield's stature as founder and first director of the Montreal Neurological Institute, a world-famous center for brain research and surgery, explains why the Soviet government sent for him when physicist Lev Landau (see Weisskopf conversation, Chapter Four) was in a coma after an auto accident in Moscow. Penfield advised not to operate and Landau recovered without surgery. While convalescing in the hospital, he received the Nobel prize in physics.

He was the son of a frontier doctor who died when Penfield was a child, his ambitious mother, a writer and Bible teacher, encouraged him to attend Princeton. Had it not been for William James, Penfield would probably have made his name in college football as a prelude to life as philosophy professor. He was a philosophy major at Princeton when he read James's *Principles of Psychology*, and set off on a quest for the secrets of the brain and mind. Like James himself, Penfield took a wide view, tackling the enduring questions: Is the mind or spirit separate from the brain? Does telepathy—communication of mind to mind—exist? Could the mind survive the death of the brain?

As a Rhodes scholar at Oxford, Penfield continued his studies under Sir William Osler and Sir Charles Sherrington, 1932 Nobel prize winner for discoveries regarding the function of the neuron. Then, after a period of animal experimentation, he became one of the world's leading neurologists, investigating what William James called "the ultimate of ultimate problems." As Penfield put it: "The time has come to look at (James's) two hypotheses, his two 'improbabilities.' Either brain action explains the mind, or we must deal with two elements."[1]

As a neurosurgeon specializing in epilepsy, Penfield reported in the 1960s that through his innovative surgery he had cured half of his over 1,000 patients and improved the conditions of another 25 percent. His method was to lightly stimulate their exposed brains with an electrode. When he reached a spot where patients said they felt as they did at the start of epileptic seizures, he removed that small area. Most patients were fully conscious under local anesthesia so Penfield could communicate with them during the operations.

He was following his probing procedure when, to his surprise, a woman patient said she heard music. Each time he stimulated the same spot, she heard the same music. To test if it was merely her imagination he tried to trick her by pretending to use the electrode, or by placing it on a different part of her brain. But then she said she heard nothing.

A 12-year-old boy cried out, "Oh gosh! There they are, my brother is there. He is aiming an air rifle at me." When his brain was stimulated again, he said, "My mother is telling my brother he has got his coat on backwards. I can just hear them." A secretary told Penfield, as he lightly touched three points on her temporal lobe that she first heard a voice from the past—a mother calling her little boy—then sensed she was near a river listening to a man and a woman calling out, and finally experienced being back in an office and aware of "a man leaning on a desk with a pencil in his hand."

This "experiential response," as Penfield called it, happened to 40 patients, as if their brains were tape recorders that could replay past events. Although no experiences involved action by the patient: none, for example, relived a time when they had been writing, eating, drinking, or lovemaking.

From his unique, firsthand observations of exposed living brains, Penfield was able to make an elementary map of the brain, linking specific areas with bodily functions. One discovery he made was that in most right-handed individuals the left side of the brain controlled their speech.

Away from his work Penfield and his family—wife, two sons, and two daughters—often headed for their farm where they skied, played tennis, and sailed on a nearby lake. He also traveled a lot to share his insight with medical groups. The Americans gave him the Medal of Freedom; the French, the Legion of Honor; and the British, the Order of Merit. In presenting him with an honorary degree, the president of Princeton spoke of this "strong and gentle man" as "a healer rather than an explorer, [who] with extraordinary dexterity penetrates the recesses of the human brain and restores lives of usefulness and happiness to those who had been facing the future without a single ray of hope."

Shortly before his death at age 85, I contacted Dr. Penfield at his farm near Magog, Quebec. My first question was: How much of the mind is still a mystery?

Penfield: The nature of the mind, absolutely. But what the mind does is not a mystery. We can see what the mind does, just as we can study what the brain is doing.

To get to your famous discovery. Can you describe it?

Penfield: It happened accidentally the first time. I was probing a woman's brain with an electrode when she said she heard a melody. I was so astonished I restimulated the same spot some 30 times. Each time she heard the same melody from the chorus to the verse and hummed along with the instruments. I got an old dictaphone to record it.

And later this happened in the same way with many other patients?

Penfield: Yes. I remember a young man who seemed to think he was at a baseball game watching a boy climbing under the fence to get in, and a mother felt she was back in her kitchen and could hear her son playing outside.

Whenever you placed the electrode in the same area of the brain?

Penfield: Yes. Always the temporal lobe.

If one wanted to write an accurate and complete autobiography from birth on, would it be feasible to get total recall of one's past life simply by touching part of the exposed brain?

Penfield: Oh, no. Where we cut in on the record of the stream of consciousness seemed to be accidental. On the other hand, there is a mechanism, in which the hippocampus is involved, which activates the record of the stream of memory. Now there is no way of starting the machine going and letting it keep up. It takes just as long for the record to redo itself as it took in the first place to lay it down. So I suppose that times of unconsciousness, as in sleep, would be blank. But it would take a lifetime and a new order of experimentation and morality [experimenting on exposed, living human brains] to completely recover the stream of memory.

Take a big leap into the future. Is it predictably possible?

Penfield: That's a question I've always wondered about. If I kept the electrode on, which I never did, would the recall of events go on and on and on? I don't know.

Why didn't you keep the electrode on?

Penfield: Because I always had a patient on the table whom I was trying to cure. They were not laboratory animals.

You also activated the motor area of the cerebral cortex that made the patients' hands move.

Penfield: Yes, and when I asked, "Why did you move your hand?" they said, "I didn't. You made me move it." And when I caused a man to vocalize and I asked him why he gave such a cry, he said, "I didn't. You pulled it out of me."

What sort of cry?

Penfield: Ahahahahahahah. That kind of sound.

Was the cry negative? Did it indicate fear?

Penfield: No. None of that, if I may say so, psychiatric nonsense.

[I had assumed the man might have been frightened of the operation and a fearful cry understandable in the circumstances.]

Do you agree that a hypnotist works on a person's mind, and the surgeon works on his brain?

Penfield: Absolutely. There has never been any evidence any time of causing a patient to think of something, or to desire something [through probing the brain]. After all the brain is a computer that we're stimulating, not the mind. We surgeons are not affecting the mind.

Is hypnosis also a mystery?

Penfield: Hypnosis and with that, suggestion, is a very important field which, as far as I'm concerned, is unexplored. I kept away from it because of all the misunderstandings and bunk that goes with it. But it's very important to know now the basis on which it depends. The brain is able to block an incoming sense of pain, and other things. The brain is much more occupied, constantly, with inhibition than it is with action. I think the medical profession should learn about and make more use of hypnosis.

What do you think of sleep learning?

Penfield: As far as I can see the talk about suggestion during sleep, or suggestion when a man is not paying attention, is rather nonsense, Because I feel that nothing is learned, nothing remembered unless attention is being paid. So that the use of suggestion during sleep should remain where it belongs: in that charming little book, *Brave New World* by Huxley, or Jules Verne's *Twenty Thousand Leagues Under the Sea*.

Dream researchers claim that some people have prophetic dreams.

Penfield: Well, I don't think we dream unless consciousness is returned.

You wouldn't describe dreams as products of the unconscious mind?

Penfield: No, not at all. Deep sleep and the unconscious are different. The unconscious is another story.

How restricted is the mind?

Penfield: From a scientific view the mind can only find expression through the brain. Now there may be extraneural communication in the way of prayer, between the mind of man and the mind of God, in the way of extrasensory perception. I don't think it should be called perception necessarily. But scientifically such communication does not exist, and I see no way of proving it scientifically.

But didn't you encourage scientific telepathy experiments?

Penfield: The Rockefeller Foundation supported J.B. Rhine's investigations for a while, and asked me if they should go on doing it. I was certainly in favor of supporting him and thought it was a very good effort of his to put telepathy to the test.

Did they continue to support him.

Penfield: Yes, for a long period, with some doubts. As far as I know, the evidence is somewhat in favor of their being some extraneural communication.

What aspect of the brain would you advise future researchers to tackle?

Penfield: The mechanism of focusing attention. There quite clearly is a mechanism within the brain that is directed by the mind. And it's that mechanism which makes it possible for the mind to program the brain. And it's then somehow or other during the focusing of attention, in the light of attention, that the engram [a permanent effect in the mind as a result of stimulation] is laid down. Nothing that goes on in the brain is remembered, that is the engram is not added, unless conscious attention is focused on it.

Wilder Penfield's eloquent conclusion about the relationship between the mind and brain is painted on a rock on his Canadian farm. Two years before he died he had painted the Greek word for "mind" or "spirit" on a large rock and connected it to the painting of a torch, symbolizing science. On the other side of the rock, Penfield painted the outline of a human head and inside it, a question mark. He joined the two paintings with a solid line to express his belief that studying the brain would ultimately solve the mystery of the mind.

Six months before he died, he changed his ideas. Braving bitter weather and wearing several sweaters, he struggled through deep snow to his painted rock. He no longer believed the brain held the secrets of the mind, or of consciousness. Now he thought it more likely that mind and brain are two distinct, separate—though linked—entities. So he changed the solid line on the rock joining brain to mind to make it an interrupted line.[2]

Had Penfield's will to believe overcome his scientific objectivity? He was, after all, raised in a devout Christian family, believing "there was work for me to do in the world and that there is a grand design in which all conscious individuals play a role."[3] It's hardly surprising he expressed his delight that as a scientist he could believe in the existence of a mind or spirit apart from the brain.

Yet he conceded the challenge remains for scientists, free of philosophical or religious bias, to prove or disprove his belief.

Chapter Twenty-One.
Roger Sperry

"[Consciousness is] one of Nature's best kept secrets."
—Roger Sperry
"Of all the things that separate men from animals, I suppose the most
obvious is reason, left brain activities."
—Roger Sperry
"I hold subjective phenomena to be primary, causally potent realities as
they are experienced subjectively, and not reducible to their
physiochemical elements."
—Roger Sperry

Like Wilder Penfield, Roger Sperry made his discoveries through epileptic patients. Sperry's subjects, however, had received more radical surgery: the nerve connections between the left and right sides of their brains being completely severed, which modified or entirely eliminated their seizures.

Sperry's major breakthrough was to establish the previously unknown function of the cluster of some 200-million nerve fibers know as the corpus callosum in the skull's center. He found that it was there to exchange information between the brain's two hemispheres. Consequently, when severed the two halves acted as separate entities.

Surprisingly, the operation did not radically change the personality; though creating two independent minds in one skull could cause odd behavior, such as a patient trying to pull down his trousers with one hand while trying to pull them up with the other. Another time this patient threatened his wife with his left hand while defending her with his right.

In testing and studying his patients, Sperry realized that the right side of their brains handled nonverbal ideas, simple words, and simple arithmetic; while the left side handled verbal ideas, sophisticated language, and complex calculations. Another outcome of the split-brain surgery was to make one side of the brain blind to what the other side saw.

Sperry also led the revolution that overthrew behaviorism, the prevailing doctrine which viewed human beings as machines and rejected the concept of the mind. "Through skillful experiments," said a former colleague, Viktor Hamburger, "Sperry was a great producer of facts about the brain, facts on which the theories of his first two teachers, Paul Weiss and Karl Lashley, were brought to grief. I know of nobody else who has disposed of cherished ideas of both his doctoral and postdoctoral sponsor, both at the time acknowledged leaders in their fields."[1]

Having successfully demolished the theory that we can be completely explained in physiochemical terms, Sperry, offered one of his own. In his writing and lectures he advocated mentalism in which "subjective phenomena, including mental images, feelings, memories and other cognitive contents of inner experience that had long been banned from scientific explanation by rigorous objective behaviorist and materialistic principles [are considered] legitimate explanatory constructs."

"The classical view," Sperry wrote, "reduces everything to physics and chemistry and ultimately to quantum mechanics or some even more elementary, unifying theory. By this long dominant physicalist–behaviorist paradigm there is no real freedom, dignity, purpose or intentionality. According to the new mentalist view, by contrast, things are controlled not only from below upward by atomic and molecular action but also from above downward by mental, social, political, and other macro properties. Primacy is given to the higher level controls rather than to the lowest."[2] And there is, according to Sperry, interaction between mind and brain during which mental states are casually influenced by brain states and vice versa.

For his work at Caltech giving "an insight into the inner world of the brain which hitherto had been almost completely hidden from us," Sperry was awarded a 1981 Nobel prize.

Roger Sperry was born in Hartford, Connecticut, and educated at Oberlin College. He earned a doctorate in zoology at the University of Chicago and held academic positions at Harvard University and the University of Chicago. Sperry went to Caltech in 1954 and was there for 30 years until he retired. He was married to his wife, Norma, for 45 years, and they had a son and daughter. Though confined to a wheelchair because of a degenerative neuromuscular disease, he was intellectully active until a few days before his death at 80 from a heart attack on April 17, 1994.

Discussing with Dr. Sperry his work and ideas I said: I read that up to the age of ten or so both halves of the brain operate. After that age, speech becomes specialized in the left side of the brain. If the child suffers brain damage on the left side then the right side will take over the language functions. But the same damage in those over ten years of age leaves them permanently mute. Is this true?

Sperry: I suppose it is in a majority statistical sense. But there are people who are ambicerebral—especially women, who tend to be more bilateral than men.

Do both halves of the brain operate independently, so that at times the left half doesn't know what the right half is thinking?

Sperry: That's largely true in split-brain patients who have had their brains surgically divided. It's not true of normal people.

You pointed out that our society was built on the verbal and rational, controlled by the left side of the brain. Do you think this is good or bad?

Sperry: It's mixed. I don't have any final statement on that. We all stress reading, writing, and arithmetic. Everything depends on perspective and contrasts and backgrounds. Of all the things that separate men from the animals, I suppose the most obvious is reason, left brain activities. Animals probably have more soul, emotion, etc., than some people.

Soul in what sense?

Sperry: The popular sense, music and so on.

What have animals to do with music?

Sperry: You've heard of soul music and soul feeling, etc. Well, the animal world has this rich emotional content, so far as one can tell from objective observation. The way a dog or cat acts or a horse, it could well have rich emotional feeling. [This is in line with Sperry's belief that "the new acceptance of consciousness along with the changed concept of the mind-brain relation applies also to the animal mind and brain with consequences for the treatment of animal awareness and behavior."[3]]

Are human brains as individual as human faces?

Sperry: Even more so, because they have much more microscopic detail, so there's much more room for variations in brain structure.

Then they're even more individual than fingerprints?

Sperry: True.

What do you think of attempts to discover signs of Einstein's genius by examining minute portions of his brain?

Sperry: In general there has been the conclusion that brain function and quality are not correlated with size or anything we can discriminate. That doesn't mean it's not there. We just haven't found it yet.

How do you work?

Sperry: I usually get in about nine, work till about twelve, then go home for lunch and hopefully a nap, come back at three till six or so. I do no teaching.

Mostly reading and thinking?

Sperry: Reading and writing. My aging brain is such that I don't rely on it too much any more. I use the new ideas and try to explain them to people who haven't caught up yet.

When do you work best?

Sperry: Early morning is the best time for thinking for me. And if I have problems that are stewing I'm apt to get answers at that time more than any other.

John Eccles was given a 1963 Nobel prize for work on the neurophysiology of the spinal cord. He thinks that the mind has a spiritual source and believes "there is a Divine Providence over and above the materialistic happenings of biological evolution." And, that "there is a fundamental mystery in my personal existence, transcending the biological account of the development of my body and my brain. That belief, of course, is in keeping with the religious concept of the soul and with its special creation by God."[4] As a teenager, apparently, he had a mystical experience which may be partly responsible for that belief. He and Wilder Penfield tend to believe that mind and brain are separate, and so the mind could possibly survive the death of the brain.

Sperry: I know that Eccles is intensely religious. But I don't agree with him. I wholly reject anything supernatural, mystical, or occult, in favor of reality validated by science. The mental qualities used to be conceived in nonphysical terms, but we now view them as emergent properties of brain processes. Everything indicates that the human mind and consciousness are inseparable attributes of an evolving, self-creating cerebral system.

George Wald is attracted by the idea that consciousness suffuses the universe. Do you have any sympathy for that view?

Sperry: Not much. It's an old idea. [Not as Wald expresses it.] It's been around a long time.

But aren't you then a strict behaviorist, insisting that mind and consciousness should be ignored and only the physical brain and physical activities be studied?

Sperry: No, I am not. Since 1965, for the past 20 odd years, I have referred to myself as a mentalist. I hold subjective phenomena to be primary, causally potent realities as they are experienced subjectively, and not reducible to their physiochemical elements. I am now interested in the implications of a new concept of consciousness. In the early 1960s, psychology was behaviorist and so was I. Five years later there was a consciousness revolution in behaviorist science, in which behaviorists were overthrown and replaced by a new mentalist or cognitive psychology. And I shifted, too, at that time to a causal-active-functional aspect of consciousness. [This meant discrediting the theories of his teachers, Paul Weiss and Karl Lashley.]

Could you explain causal-active-functional?

Sperry: Science previously had always assumed that brain function was complete in itself. There was no place in the neuroscience picture for any causal influence of consciousness or mental subjective states. These were assumed to be parallel epiphenomena, an aspect of brain function but not causative.

Is it accurate to say that consciousness or the mind is in a superior position to the brain and the mind causes thoughts?

~~Sperry: Well, it applies to the revised view.~~

Which you favor.

Sperry: Yes. By 1970, the whole field of psychology had switched from behaviorist to mental-cognitive. The main fact that caused the overthrow of behaviorism was the concept of consciousness as causal rather than as passive or noncausal. Behaviorism could cope with and coexist with all the other developments, computers and cognitive science and so on. But it could not coexist with the concept of consciousness as causal. So I think that's quite a step forward, to change the paradigm of a whole discipline of science.

Though consciousness is still a tremendous mystery, isn't it?

Sperry: I'd say it's one of Nature's best kept secrets still. But there's a qualification there. I suspect that the issue of whether consciousness, whatever it is, is causal or noncausal is maybe more critical than the secret of how you build consciousness into a computer. I suspect that if hummingbirds and insects can be conscious, as many people agree, then the secret programming or processing is not all that complicated. Once we've discovered it, it will seem a kind of simple thing. And it doesn't make all that difference. But the issue of whether it's causal or noncausal has tremendous implications.

What's one tremendous implication?

Sperry: Science by the traditional noncausal view eliminates all subjective human experience from the world view. It erases it, says it's not a causal reality. You've heard of the two cultures of C. P. Snow, and that scientists and humanists don't get along. Well, a lot of that was attributed by Snow to language and subject matter. But behind it all, most importantly was the conflicting world view in which human subjective experience seemed important to the humanists and totally unimportant to science. Now this new view acknowledges the subjective aspect of an experience and bridges over that conflict of cultures.

Does your new world view assume that, ultimately, scientists will be able to examine subjective experiences?

Sperry: Yes. They are still brain processes. They are emergent properties of the brain processes.

If you suspect the brain itself is the source of consciousness, how close are you to finding out how it works?

Sperry: Every time we make a discovery it opens up several more problems. We're a long way off.

What do you think is the purpose of life?

Sperry: The answer to that depends on whether you believe consciousness to be mortal or immortal, reincarnate or cosmic, and whether you think it's brain-bound or universal.

Which do you believe?

Sperry: My point is that if consciousness is mortal then your values are very different from those if it's immortal.

Wilder Penfield concluded it may be immortal.

Sperry: And lots of theologians. But I don't see any convincing evidence anywhere that consciousness can exist apart, separated from the living, active brain.

So what in your view is the meaning or purpose of life?

Sperry: Just that the body and the juicy nerve cells and their animalistic appetites, etc. and properties of consciousness, don't survive. And most religions agree with that. But with the new mentalist interpretation of consciousness, if you ask, "What is best? What would you most like to preserve?" then it's not the daily appetites and aspirations, etc. It's the very peak achievements of the mind, and those are, in a way, expressed in Beethoven, Shakespeare, Michelangelo, etc., etc. You can say that their best is still with us. Their bodies, their lower levels of conscious experience are not, but the very highest aspects of them as persons, you could say, still survive in a sense. Of course, we can't all be Michelangelos and Beethovens, but the principle still applies. A few things survive beyond the brain—paintings, ideas.

Do you believe in a Supreme Being?

Sperry: Not if you define a Supreme Being as a person, an intellect apart from creation itself, a designer. I see the supreme plan for existence intrinsic in the evolving nature, in the evolutionary process itself. Creation cannot be separated from the creator.

When you say that you talk as if there is a creator.

Sperry: I think we are here because forces have made and moved the universe, and created man. But these forces are not personal. They're the forces recognized by science as being operated in evolution.

But isn't the origin of those forces a mystery?

Sperry: Well, it's certainly not comprehended yet. But it appears to evolve in simple steps at first and gradually get more and more complex. So the "creator," to use that term in quotes, is self-evolved, self-determined, not a prescient of some divine intellect.

Your version sounds even more mysterious to me.

Sperry: Just as respectable. In the eyes of science, put simply, the creator is the vast interwoven fabric of all evolving nature, a tremendously complex concept. It includes all the immutable and emergent forces of cosmic causation that control everything from high-energy subnuclear particles to galaxies, as well as casual properties that govern an individual's brain function and behavior. The idea evolves as science evolves.

It still sounds like a big mystery.

Sperry: It is.

Did you have a religious upbringing?

Sperry: Very little. Some Episcopalian.

How did you lose what little faith you had?

Sperry: My opinion is, that thanks to Freud, with assistance from astrophysics, science can be accused of having deprived thinking man of a Father in heaven, along with heaven itself.

Have you read much of William James?

Sperry: I was kind of reared on him.

Then you know how interested he was in psychical research.

Sperry: I hadn't realized that.

He spent years investigating Mrs. Piper, a famous Boston so-called medium, among others. He concluded his tests frustrated, but still open-minded.

Sperry: It's this kind of belief and attitude that has gotten us into the mess we are in today: the churches and so on.

Do you think there may be intelligent life on other planets?

Sperry: What I think of that subject would be of no consequence.

One reason I speak with people like you is that not only are you experts in your fields, but you probably have wider interests and more contacts with experts in other fields than the average person.

Sperry: I would not be surprised if they discovered another planet that had conditions in which life had evolved.

So you're open-minded about that, but not about religion.

Sperry: My position is that if you once believed in that sort of thing it leads to values that perpetuate the wrong efforts and policies we've been living under. But it's not a matter of what you ultimately believe. It's a practical matter of what we can risk these days in our beliefs. We need a new perspective, new values, new beliefs, if we're going to do anything about our population problem, etc., etc. If you take the centralist view of consciousness, the new paradigm of science that is supported by the behavioral, evolutionary, social, humanistic sciences now, and is spreading; if you take this paradigm of science you come out with beliefs and principles and policies that would make for survival.

Finally, Dr. Sperry, how much of the mind is still a mystery?

Sperry: All of it. The centermost processes of the brain with which consciousness is presumably associated are simply not understood. They are so far beyond our comprehension that no one I know of has been able to imagine their nature.

Chapter Twenty-Two.
Torsten Wiesel

*"The eye and the brain are not like a fax machine, nor are there little
people looking at the images coming in."*
—Torsten Wiesel
*"If you do not operate on a child, particularly one with monocular
cataract before probably half a year, the child will never regain
full vision in that eye."*
—Torsten Wiesel

As a boy Torsten Wiesel, the youngest of five children, lived on
the grounds of a mental asylum in Uppsala, Sweden, where his father was in charge and chief psychiatrist. By the time Wiesel was 30
he, too, was a psychiatrist, working with children, and a year later
a fellow in ophthalmology at Johns Hopkins University Medical
School. At 36 he moved to Harvard Medical School as assistant professor of neurophysiology and neuropharmocology, and at 45, in
1973, became chairman of the department of neurobiology. From
1978 to 1979 he was president of the Society for Neuroscience. In
1981 he and his partner David Hubel shared the Nobel prize for
medicine with Roger Sperry. Two years later Wiesel was appointed
head of the laboratory of neurobiology at Rockefeller University.
Since 1992 Dr. Wiesel has been president of the university. A keen

photographer and art collector—from pre-Colombian to modern—Torsten Wiesel has an easy, friendly manner, speaks English fluently with a slight Swedish accent, is in his early 70s and has been twice married and divorced.

Day after day in the summer of 1958, Torsten Wiesel and his Canadian partner David Hubel worked in the windowless basement lab at the Johns Hopkins Medical School. Using fine-tipped micro-electrodes to pick out individual brains cells, they hoped to begin to map the cerebral cortex of the mammalian brain, by first recording the response of one brain cell to patterns of light. But the brain cell kept its secret.

The mammal they chose for this experiment was an anesthetized cat, its eyes held open and an electrode implanted in its visual cortex. When they stimulated retinal cells with light they, as expected, burst into life. But cells in the visual cortex—the brain area behind the back base of the skull to which retinal neurons project via optic nerve and thalamus—did not respond.

The two scientists who had become friends kept at it, working into the night but getting nowhere. Time after time they shone a bright spot and then a dark spot on different areas of the cat's retina, projecting the bright spot through a hole in a brass rectangle and the dark spot through a black spot glued to a glass microscope slide.

Then after four frustrating hours they got a few sputtering responses from the brain cell. When it happened the slide projector was pointed at the retina's midperiphery. So they decided to focus on this area, thinking that it might be where the retinal region fed into the cortex.

They were placing the black-dotted slide into the projector slot, when it stuck. So they unjammed it and began slowly easing it back in. As they did, the edge of the glass slide threw an angled shadow on the retina. And that's when the brain cell responded—sounding through an audiomonitor like rapid gunfire.

"The door to all the secrets," as Wiesel put it later, had been opened by accident. Others have described it as breaking the code of how information from the eye is decoded by the brain.

According to Wiesel this chance discovery of brain cell activity only occurred because the edge of the glass slide had cast an edge of the shadow on the retina at a particular angle.

What is now known as "orientation-selectivity" is regarded by many as the most remarkable discovery about cells in the cerebral cortex. Subsequent Wiesel–Hubel research revealed how information in the individual's system is coded by nerve cells—the language that is used by the brain to communicate the information.

This won Torsten Wiesel and David Hubel the 1981 Nobel Prize in Medicine, which they shared with Roger Sperry.

Dr. Wiesel spoke to me from his office in Rockefeller University on January 18, 1995.

Brian: You grew up living on the grounds of a Swedish mental hospital where your father was in charge of the place and its chief psychiatrist. Wasn't it frightening?

Wiesel: No, I felt more secure within the hospital than outside.

The patients weren't dangerous?

Wiesel: Some were violent. But most of the patients let out on the grounds were not considered to be dangerous. I used to play soccer with them.

When I was seven I lived opposite a mental asylum in Wales and it was frightening especially when the warning siren went off to say a patient had escaped.

Wiesel: Sometimes they came into the house and were angry at my father. I remember the only thing I was worried about was when Lindbergh's son was kidnapped [in the United States] and we were worried that some kidnapper would come through the windows. But we didn't associate that fear with any patient.

You described yourself as "a mischievous student, rather lazy, interested mainly in sports." How were you mischievous?

Wiesel: Probably during lessons. I had a restlessness perhaps, but I still think my classmates liked me. When I began school I wanted to establish myself, so apparently I had fights with everybody in my own class and the class above and the class above that. But I never became a bully.

What age did that start?

Wiesel: I was about six.

Did any novels have a great effect on you?

Wiesel: I read a lot and collected books as a young man. I liked particularly poetry, and probably read more poetry than novels. My favorite poet, Gunnar Ekelof, had the abilty to express a mixture of sensations, abstract ideas, and complex thoughts in a beautiful way while remaining comprehensible. I look at poetry as similar to chamber music—a form that has its own harmonies and rhythms and that gives rise to fantasy and associations in the mind. Ekelof stimulated that in me.

Are any of your two brothers and two sisters scientists?

Wiesel: No. I also have two half brothers and a half sister. One of the half brothers is head of psychiatry in Uppsala University. One brother is a lawyer and one is a medical doctor.

Did your father encourage you to become a psychiatrist?

Wiesel: No. I had a brother who became schizophrenic and then lived in a mental hospital, and as a medical student I worked in the mental hospital as a night nurse to pay for my studies. Then I became very interested in neurophysiology, because I had a very good professor there who also stimulated my interest. So after that I had a divided attention to neuroscience and psychiatry. After some years I decided that our knowledge about the brain was such that I wanted to learn how it works.

How is your schizophrenic brother?

Wiesel: He was hospitalized for nearly a decade. This was before the pharmaceutical revolution. Then he could be released and he's pretty good. He's still alive and doing well, relatively speaking. And he's completely drug-free.

Is schizophrenia still a valid diagnosis?

Wiesel: If you have hallucinations and paranoia then the symptoms fall . . . You know, either you're manic-depressive or schizophrenic. Then you have the whole range of psychopathic behavior.

It's broadly manic-depression or schizophrenia is it?

Wiesel: Or genetic factors involved.

Was your brother's assumed to be genetic?

Wiesel: It's in the family. The family's probably a little-bit crazy.

Doctors are supposed to go in for the branch of medicine that most applies to them. So some cynics say psychiatrists must be a bit crazy. Is it a myth to say more Swedes commit suicide per capita than inhabitants of other western countries?

Wiesel: It's difficult to know, because Swedish registration is better than in most other countries, and it's not a crime in Sweden to commit suicide. So doctors report it, but in Catholic countries they don't. That probably distorts the statistics. I had another brother who was a painter. So there probably is that element in the family of people who are not conformists.

You can't call people crazy because they're artists.

Wiesel: No, no. But misbehavior when you're mentally ill and when you're nonconformist means you can't integrate yourself into society so easily.

Was your major discovery that if the brain is visually deprived in early life it never develops its full potential?

Wiesel: Yes, that would be the conclusion. But we started out studying the normal adult, and subsequently when we knew something about how the visual cortex works, we then asked the nature/nurture questions.

What impetus drove you to do research that concluded that early deprivation would stultify brain development?

Wiesel: Both my partner David Hubel and I had a medical background and knew from the clinical literature that kids with cataracts never regained full vision. Usually, in those days kids were operated on for cataracts when they were five or seven, later than they are now. There's a long, interesting literature on the subject. And these were behavioral experiments done on raising monkeys in the dark. [In experiments at Harvard Medical School, Wiesel and Hubel found that if one eye of a kitten was closed for four days, detectors in the brain corresponding to that eye did not recover.] Philosophically it's always been interesting to get hold of nature/nurture questions. We thought this would be a very direct experiment to do. So we closed one eye and said we could compare the two eyes and get a very clear answer.

Which was?

Wiesel: That the visual cortex is to a great extent wired up and ready for use at birth. But we found that mammals deprived of early visual stimulation had dramatic changes in the structure of their visual cortex. To ensure the proper development of the visual system there's a critical period early in visual development—a matter of weeks—in which innate neural wiring and visual experience must interact.

Would the same apply to lack of tactile and aural stimulation?

Wiesel: It would apply to all sensory systems, I'm sure.

What evidence do you have for your belief that "throughout life, experience continues to modulate the fine patterns of cortical connections, allowing us to acquire new skills and knowledge," as you wrote in *Science* [June 17, 1994].

Wiesel: I obviously have no direct evidence from human experiments, but I assume that since we can learn to recognize new faces and acquire new languages it must mean the establishment of new patterns of neuronal connections. In terms of animal experiments, my colleague Charles Gilbert and I have shown that connections in the visual cortex can be modified by a specific experimental paradigm.

What did you discover about human infants born with cataracts?

Wiesel: That if you do not operate on a child, particularly one with monocular cataract before probably half a year, the child will never regain full vision in that eye. I hear from my ophthalmolic colleagues that you should operate on a monocular cataract within the first few weeks. If both eyes have cataracts you can wait a little bit longer, because of the competition between the two eyes, presumably.

Do all surgeons in the U.S. now act on your discovery and operate early?

Wiesel: It took 20 years, but I think now most ophthalmologists are taught about it in medical school, and in some clinics they screen all newborn babies for cataracts.

Did you read *The Bell Curve* by Charles Murray and Richard Herrnstein?

Wiesel: I have the book and I looked at it. It's not a very good book. The only time I agreed with Newt Gingrich was his comment on *The Bell Curve*, when he said it was a lot of nonsense. It's not really scientific in a sense. They use material as it suits their agenda. I was very perturbed because the book argues that a child born with a low cognitive ability cannot raise its intelligence by outside intervention. I believe that David Hubel and I showed very clearly that nature and nurture both play an important role in brain formation and that should be so in terms of intelligence, emotion, and socialization. But our experiments led us to conclude that to prevent damage to the system one should act when the child is very young.

You apparently agree with social scientists who believe that early attempts to improve the environment of disadvantaged children can also enhance their mental development.

Wiesel: Yes. I have concluded that a balance between the environment and heredity, nature and nurture, have an effect on the operations of the brain.

I take it from your comment about Gingrich that you're a liberal.

Wiesel: I'm Swedish, and I suppose I was brought up in a society where certain values were socialistic.

Are we aware of things because of the retinal cells reacting or the visual cortex reacting?

Wiesel: The eye and the brain are not like a fax machine, nor are there little people looking at the images coming in. Both eye and brain have to be active. The image that falls on your photoreceptors is being decomposed into critical elements represented in terms of form and shape and color. There are about 200 million photoreceptors and one million ganglion cells going into the brain. So that means we have to reduce information and optimize the information in some fashion.

And as you project into the brain there are again several hundred million nerve cells in the visual cortex that can then analyze the information again and distribute it to higher centers where you will finally have a perceived image. We don't really understand what happens. It's still a mystery. Our main problem—to understand the neural basis of perception—is still to be made.

Is the old saying that we lose millions of brain cells every day true?

Wiesel: I don't think so. It's greatly overstated. We probably lose a few. We don't have any new brain cells made, so we have to live with what we have. My philosophy has always been to take good care of what you have.

Did you see the recent suggestion that song birds may create new brain cells, memory cells, to recall where they put hidden seeds?

Wiesel: Yes. That's down here in Fernando Nottebohm's laboratory at Rockefeller University. [Dr. Nottebohm and his colleague Dr. Anat Barnea, now of Tel Aviv University, discovered that each October, before winter sets in, the hippocampus of the brains of black-capped chickadees is considerably enlarged. The hippocampus is associated with memory.] Not only song birds, but also a recent study here at the Rockefeller showed new brain cells are made in the mouse.

But there's no hope for mankind?

Wiesel: There's no evidence for that. But the people doing this work with birds believe we may be able to develop ways—if it does not happen in our brains normally, they are not sure that's the case—but with the discovery of the various growth factors in the brain we may come to the point where you actually could stimulate and make new nerve cells that could migrate. Then you have the problem of how would these cells know how to migrate to the right place? There are a whole set of very interesting questions. That would be the ultimate triumph for science—to keep on having your brain always young.

What have you worked on since you ended your partnership with David Hubel in 1976?

Wiesel: I worked with another colleague, Charles Gilbert, and we've made some discoveries which probably wouldn't win another Nobel prize, but still are of some interest in the field in terms of how a new pathway within the cortex makes cells interconnect with each other and speak to each other.

George Wald says color-blindness could never be cured. Do you agree?

Wiesel: If I was color-blind, I would not necessarily want to take the risk of having it cured. What's happening now, which is of interest to people with retinal defects or where the retina is dying, is that animals begin to replace the photoreceptors [light-sensitive cells], so that you can transplant new cells into the eyes. So it could be possible to restore vision in some cases. But the cones or color cells are primarily in our fovea, a very small area half a millimeter big, and if you burn that out, for example, you cannot read any longer. And most of our color cells are in that area.

When we see things in our dreams what are we using to see?

Wiesel: There's been a lot of speculation about that. It's still a speculative area. My own belief is that there's some evidence from studies using brain imaging that when you perceive something, both the primary and high visual areas are active, not just one small area. So when you dream, presumably you activate a set of these memory circuits. Sometimes the dreams are unreal, mixing things up.

Crick writes that "emotional memories involving fear are permanently engrained in the brain. They can be suppressed but never erased."

Wiesel: If that's his experience. Do you think it could be true?

It makes sense.

Wiesel: Yet there's probably a lot of emotional stuff that you've forgotten.

You remember when Wilder Penfield was probing the brains of living people and it seemed to elicit their memories? And he speculated that if he'd kept doing it they'd recall everything that had ever happened to them?

Wiesel: Yes. That turned out to be overstated.

You don't think our brains holds the record of our entire lives that could be recalled under hypnosis?

Wiesel: I wouldn't say so. The best way for me to remember things is in an environment. Say I go back to Sweden, then I remember things better when I meet people—associative memories.

If you were a Swedish novelist wanting to use a Swedish background in your novel some people would advise you to go to another country where—perhaps because you can't see and hear what you want to describe—you're forced to recall it, and then the most vivid memories would come to mind. Here's something else Crick says about the brain: "You, your joys and sorrows, your memories and your ambitions, your sense of personal identity and free will are in fact no more than the behavior of a vast assembly of nerve cells and their associated molecules."

Wiesel: What else could it be? Although Jack Eccles [neurobiologist who won 1963 Nobel prize] thinks the mind and brain are separate.

Did he ever tell you about the supernatural experience he had as a teenager that made him think the mind was a separate entity?

Wiesel: No.

Apparently he doesn't talk about it. Although he told me he was impressed with Dr. J.B. Rhine's investigation at Duke University of extrasensory perception. Do those who hope to find an explanation for Einstein's genius by examining his brain, stand a reasonable chance?

Wiesel: I don't think so. Kant's brain was very small, only 1,200 grams apparently. The standard brain weight is about 1,350. And he was pretty smart.

Einstein liked him, too. Dr. Marian Diamond's report that Einstein's brain contained more glial cells per neuron in all four areas compared with the brains of eleven so-called normal males aged 47 to 80. Does that mean anything?

Wiesel: No. You know, you have to take scientists with a grain of salt in these matters. [He laughed.]

Are there any observable differences between the male and female brain?

Wiesel: That's a controversial issue. There is some evidence of differences. This came up in a discussion about the gay factor. You can see the size of a part of the midbrain is slightly different in the male and female brain. It was also found in some gay males, where it was smaller, more like a woman's. So there certainly are differences. And presumably the hormonal influences play a role.

But do you know what these differences mean?

Wiesel: The behavior is quite different between men and women—what we like, how we dress, our movements. There are many differences.

And you think that's brain-directed?

Wiesel: What else would it be?

Custom, habit, training. Men dressed more flamboyantly than women in Elizabethan days, earrings included.

Wiesel: No. I think that's why the gay issue is interesting, because a man who has a female brain behaves as a female. If you talk to a gay they say they instinctively wanted to dress like a girl.

How about masculine gays?

Wiesel: That wasn't brought up, if it was a masculine or feminine homosexual. It's interesting. It's not my prime interest in life, but it seems obvious that with the very striking physical differences we see between male and female, these are also going to be reflected in their brains.

Would you also find it in animals?

Wiesel: Of course. In birds it's only the male that sings. The male canary has to sing for his woman.

Why does he need new brain cells to do that?

Wiesel: Actually, female birds also generate new brain cells each year. Why birds generate new neurons is unknown. One possibility is that many birds learn new songs each year, which requires new brain resources; yet because it must fly, the size of a bird's brain cannot keep getting larger, as can its terrestial counterparts—an animal with a monkey-sized brain, for instance, would never get airborn. Thus birds use the strategy of brain death and renewal to meet new neural demands. A male canary that learns a new song in the spring may generate new neurons to control this song-circuit; a female canary may generate new neurons to learn to *perceive* these new songs.

Would there be differences in the brains of male and female cats?

Wiesel: I would guess so. I don't know if anyone has studied that. After all, if you have two Xs and you have an X and a Y that's a lot of genetic differences.

Richard Feynman said, "The phosphorus in both brains of rats and humans is not the same as it was two weeks ago; it means the atoms in the brain are being replaced." Is that true?

Wiesel: I don't think that's very important. It doesn't do anything for me.

Do you know Gerald Edelman?

Wiesel: I know Gerry, yes.

He got his Nobel prize for establishing that the immune system works on the Darwinian principle; the body recalls the invading virus and fights more ferociously in the second attack. Do you agree with him in interpreting brain activity in Darwinian terms? That the most used or useful areas of the brain strengthen and prosper?

Wiesel: I know his books. [He chuckled.] That's one of the nice things about scientists and ideas—that people have different points of view.

What's your response to his point about the most used or useful areas are strengthened?

Wiesel: We know from our deprivation experiment that if you deprive one eye then the other takes over more of the brain. So, some of the ideas of Gerry came from those kinds of experiments. You can express it in Darwinian terms if you want, but I don't know how far you can go down that road.

George Wald says, "Since the presence or absence of consciousness defies all scientific approach we have no way of locating it, but more, just as the electron does not have a precise position and motion, I believe that consciousness has no location."

Wiesel: Yes, I like that. I think George has a nice, subtle mind and as I mentioned, some of the imaging suggests we should not look at consciousness in terms of location.* It's an emerging property of activity that we may never be able to localize. George is very poetic, and takes a great interest in consciousness and the mystery of it. And, as you know, Francis Crick says, "Now is the time we should study consciousness."

He also says that scientists' understanding of how learning and memory occur in any species is primitive and incomplete.

Wiesel: I agree. But people are working on it. You know we are at a very primitive stage. I sometimes say that our knowledge about the brain is roughly at the stage where our knowledge of the universe was at the time of Galileo and Kepler. We're at the surface of understanding and we should be humble about it. We can't right now do what Francis wants us to do—design an experiment to study consciousness. Being sort of pragmatic and reductionist as a scientist, I believe that we should first learn more about how each part of the brain works and how they interact—for example, learn more about perception, about motor control. All the workings of the brain are more than consciousness. And out of those discoveries, which we cannot now conceive, may come an understanding of consciousness.

What area of neuroscience would you recommend to a young neuroscientist just starting out?

Wiesel: The neural basis of any perception is an interesting problem. Also, I think, memory. Neuroscience has fallen into two groups: the systems types, people doing the kind of work we do, who are interested in how complex systems work and communicate. And that sort of links up with computational neuroscience and cognitive science and psychophysics. That's an area still not very well developed. Then you have molecular neurobiology and developmental neurobiology, where you have very concrete experiments to do, to study how the

---

* Dr. Wiesel worked with Amiram Grinvald on applying "optical imaging" to the visual cortex in which the activity of large regions of the cortex are pictured on a video screen.

brain develops under the influence of growth factors and other molecular determinants. Then, of course, if you're interested in clinical aspects of brain function you have human genetics and all the different ways of studying the brain by using mutations that lead to disease—that also may lead to an understanding of the brain. Which of the areas you would go into depends on your background and inclination.

Wilder Penfield said he'd recommend the focusing of attention.

Wiesel: Again, like consciousness, that's a complicated subject.

Have you done any research into hypnosis?

Wiesel: No. I was trained in psychiatry in Sweden. Hypnosis was never taken all that seriously in the scientific world, at least not in Sweden. When I was practicing medicine there were cases I had with headaches that did not seem to respond and hypnosis was tried. It is an interesting phenomenon, but in the last 40 years I have done animal experiments, so I have no clinical practice or experience except for my young days. So hypnosis remains an interesting phenomenon as it is for you. I can hypnotize. It's not very difficult to put somebody under hypnosis. Why do you ask?

I find it fascinating. I've interviewed experimental hypnotists and watched one at work.

Wiesel: How does that relate to my life?

You're a scientist who's interested in the brain, and hypnosis has an unusual effect on the brain. I know someone who had a serious dental operation and they were told they'd have to take strong painkillers afterwards. Instead, they were hypnotized and suffered no pain.

Wiesel: Yes, there are all kinds of stories. It's like acupuncture. Another phenomenon we don't understand very well, which perhaps should be studied more vigorously.

Richard Feynman volunteered to be hypnotized when a student at Princeton because he volunteered for everything. One thing intrigued him was the post-hypnotic suggestion he was unable to resist—an interesting aspect of hypnosis.

Wiesel: Which was to find a beautiful girl or what?

He didn't need hypnosis for that. It was his normal operating procedure. Does it mean anything to you that so-called multiple personalities show different brain waves for their different personalities?

Wiesel: There are abundant descriptions of people with multiple personalities. You yourself probably have more than one personality, but may not be so extreme.

What other interests do you have?

Wiesel: Photography, African, pre-Colombian and contemporary art. In my offices are a large number of different things.

Ever thought of writing a novel or a play?

Wiesel: I don't know that I've got anything interesting to say. Maybe with increased senility [Weisel is 70] I will think I have.

How about life in a mental hospital through a child's eyes? It could be interesting to see how that shaped your way of looking at things. My uncle in England played bowls against a team of patients at Broadmoor, a criminal mental asylum, and had no trouble. [He later found that his completely silent opponent for one game had murdered two girls, and was told that several of the nurses had been savagely attacked by patients.]

Wiesel: It's fear that makes people deal with criminals and mentally ill people so badly. I have a certain compassion for them because I look upon criminals as people who are mentally ill.

That's rather the way Einstein looked on them. Did your parents know of your success?

Wiesel: My mother did not but she loved me anyway. It wasn't important to her. My father died in 1973 when I was a professor at Harvard, so he realized I'd been successful in America.

Did your father consider he had been a success?

Wiesel: He got married and had five kids first, including me, and then divorced and had three more. He told me his real ambition earlier on was to do neuroscience—but he never realized it.

Chapter Twenty-Three.
Update

Linus Pauling lived alone at Big Sur but he did not die alone on August 19, 1994. His daughter, Linda, and two of his sons, Crellin and Linus Jr. were with him.

"He had very few and mild painkillers," said his secretary and assistant Dorothy Munro, "because he wanted to study his own dying as a scientist. Even though he knew he was dying he kept his sense of humor."[1]

His doctor agreed: "To the end, it was almost as if his diseases were just another opportunity to do research. He remained active in experimental therapies, almost as a grand experiment. Up until the last few weeks of his life, his mental capabilities were just incredible."[2]

"Six hundred attended his memorial service at Stanford Memorial Church, where Stanford President Gerhard Casper told how on first meeting Pauling he was "totally awed and totally enchanted" by Pauling's unpretentious and kindly attitude. Chemistry professor James P. Collman said, "He thought big thoughts about everything. There is probably no important idea in chemistry that he didn't think about and publish about. He was often right, and often years ahead of the experimental techniques that would prove him right."

And Linda Pauling Kamb recalled how her father loved the movie "My Fair Lady," and could recite "The Walrus and the Carpenter."[3]

Pauling had taken huge daily doses of vitamin C (18,000 mg) since the mid-1970s, [the benefits are still unproven], scornfully dismissing the government's recommended daily intake for adults of 60 milligrams, because it "only keeps you from dying of scurvy."[4]

This explains the *Denver Post* cartoon by Mike Keefe. Pauling stands at the gates of heaven. A surprised Saint Peter asks, "What kept you so long? And the 93-year-old scientist replies, "Vitamin C."[5]

The legal case brought against him shortly before he died was thrown out of court.

HOW OLD IS THE UNIVERSE?

During my conversations with Dr. Jastrow early in 1994 he estimated the age of the universe as 15 billion years. In October, 1994, Dr. Wendy L. Freedman, heading an international team of astronomers, announced that her estimate was between 8 and 12 billion years. She cautioned that the estimate was tentative and would need to be confirmed by many more measurements. But her team had used the recently repaired Hubble Space Telescope to achieve the most precise measurement ever between Earth and a distant galaxy.

Who eventually proves to be right depends on the value of the Hubble constant, a figure used to determine the expansion rate of the universe. A high figure for the Hubble constant will support the Freedman team, a low figure indicating a slower expansion rate and older universe will support Jastrow.

Jastrow suggested that the best astronomer to respond to the Freedman report was his friend and colleague Dr. Allan Sandage at the Carnegie Observatories in Pasadena.

Sandage learned astronomy from Edwin P. Hubble who, in 1929, discovered that the universe is expanding uniformly and in all directions. The Hubble Space Telescope and the Hubble constant are named in his honor.

Sandage has spent his working life checking the expansion rate and consistently gets a low number, implying a 15-billion-year-old universe rather than the youngster of 8 to 12 billion suggested by the Freedman team.

I spoke with Sandage on December 7, 1994, when he first explained the Hubble constant.

Sandage: The Hubble constant is the velocity of recession divided by the distance of the galaxy from us. The red shift. When I say from us, that does not mean we are the center of the expansion. If you put yourself any place else in the universe you would also believe that everything was expanding away from you. It's the ultimate Copernican principle: everything looks like it's at the center of the expansion.

What did you think of the claim of Dr. Freedman and her team that the universe may be as young as 8 billion years but not older than 12?

Sandage: I believe they are wrong and I believe in the numbers we've gotten for many years—the age [of the universe] being 15 billion years, taking one billion years to gestate the galaxies. The age of the oldest galaxies then being 14 billion. That has come from studies of the age of globular clusters in our own galaxy and that's believed to be a quite secure value. We also believe the Hubble constant is not 80 [the Freedman figure] but 50, which is then consistent with the long age. The Hubble constant itself, if there's no slowdown of the universe's expansion, would give 19.5 billion. The difference between that and 15 billion takes account of the deceleration of the expansion over the entire age. So I can tell you, I firmly believe in the long time scale.

Dr. Duccio Macchetto, an astronomer with the European Space Agency and the Space Telescope Science Institute in Baltimore mentioned 12 billion recently.

Sandage: They have now observed galaxies which are 12 billion years old. These are not the oldest galaxies. These are galaxies in their early time of evolution, but Macchetto is not claiming that they are at the very begining of their lifetime.

You mean that the Hubble telescope hasn't penetrated to the galaxies 15-billion light years from us?

Sandage: You don't see anything there, because matter has not yet condensed.

This then confirms the Big-Bang theory?

Sandage: Oh, yes. Absolutely. It means that if one really sees a difference in the form of the galaxies back in time, you are seeing galaxies actually evolving. Now the Big Bang requires that. Whereas with steady state, if the universe had been around for ever you would not expect to see galaxies changing their appearance as you look back in time.

Dr. Alan Dressler, an astrophysicist at the Carnegie Observatories in Pasadena, made an interesting comment: "It seems that almost as soon as nature builds spiral galaxies it begins tearing them apart."

Sandage: That's not correct, because you have many spiral galaxies in our neighborhood that are as old as any galaxy. So what Dressler is saying is that once galaxies are in clusters it's a very harsh environment. But all galaxies are not in clusters. There are many field galaxies. Sixty percent of all galaxies in the local neighborhood are spirals. And we do not live in a cluster. If we were living in a dense cluster of galaxies there wouldn't be many spirals. So that's a very strong overstatement of Dressler's.

Why is Macchetto's report considered exciting news if he's merely confirming the conventional wisdom?

Sandage: It is the conventional wisdom. But the search for the appearance of galaxies when they're only several billion years old had been going on for 15 years, and had not been successful up to this point. So we've never been able to look out that far in space, and therefore back that far in time. And the appearance of these galaxies which Macchetto believes are very distant, that appearance is unusual. So it does confirm the prediction of the standard model. But the standard model has so many models it's not a surprise. As I said, up to now the telescopes have not been able to see that far out in space.

Reporter John Noble Wilford writes of the repaired Hubble telescope: "It's like being able to see noses and eyebrows on what were previously blank faces." Sounds pretty good.

Sandage chuckled: No, it's not that pretty good.

That's for newspaper reporters?

Sandage: Sure.

What would you like the next discovery to be?

Sandage: The tremendous controversy is the scale of the universe, which is the Hubble constant. The report of last month claimed the ages are half of the conventional wisdom. If that's true we don't know how to age-date the stars, because the stars cannot be older than the universe itself. It's been our contention all along that people who want a short age from the expansion rate, using their value of the Hubble constant, don't have the right number. The numbers Macchetto has quoted are our values for the Hubble constant and the age—the 14-billion-year age is the value we've gotten over the years from age-dating the globular clusters. Now if you want a much younger universe, then our age at 14 billion for the galaxies must be wrong. I don't believe that. So the two numbers, the age of the globular clusters and the Hubble constant are very strongly linked. And the so-called claim last month that the Hubble constant has been determined, and it is consistent with a short age, puts a tremendous spanner wrench in the standard model. I believe that the report is incorrect.

What's the strongest point in their argument?

Sandage: They have a distance to a single galaxy that they then identify, incorrectly we believe, with the distance to the core of the nearest great cluster, which is the Virgo cluster. That is an assumption. There's no way to prove it. They also do not use the correct value for the cosmic expansion of the cluster. It's a complicated technical problem. They use a value which is 19 percent too high.

Is that a mistake or an arbitary choice?

Sandage: It's a choice that has to be made as to how chaotic the local expansion field is and what then is the correct value, corrected for all random motions of the core of the Virgo cluster. We believe that it's been very accurately determined. They use another number. So there's disagreement at every stage.

Makes it interesting.

Sandage: [laughs] No it doesn't.

Why not?

Sandage: [still laughing] I've been working on this problem since 1953. Forty one years. It's not funny.

Chapter Twenty-Four:
Epilogue

We are part of an accident still in progress though, strangely, we don't sense it. Someone or something started an explosion and even as the pieces fly apart detectives on one of them are trying to figure out exactly what happened.

At one end of the room they've got promising leads. They've traced the explosion back to a fraction of a second after it happened and can predict the outcome fairly confidently: the explosion never ends though life does as the universe freezes. They agree that no one is to blame for the explosion. It was an accident that happened to something hot and dense and so small you might even call it "nothing." Though what caused this "nothing" to exist in the first place is anybody's guess.

As the billions of pieces flew apart and cooled, on at least one of them, water, oxygen, fish, trees, lions, hummingbirds, and humans developed, following the laws of nature. These laws were discovered and confirmed by math majors or otherwise gifted humans using informed guesses they called intuition.

Who made the laws? Nature made the laws. Why is Nature capable of making laws? It just is. Whys are questions that strike scientists dumb—even Einstein—or which they call dumb questions perhaps because they can't answer them.

Why the explosion? Why the hummingbirds? They don't pretend to know or expect to find out.

At the other end of the room a smaller group of detectives generally agrees with the Big-Bang theory but does not think it was an accident. It's also nonsense, they say, to suggest the universe was made from "nothing." It makes more sense to them or meets their emotional needs to see some purpose in the world and the immense ongoing explosion as part of a plan.

A substantial minority still put their money on the Big Crunch, with the universe eventually returning to its birthplace. If you were making a film of it you'd just reverse the movie. For this to occur there must be much more matter in the universe than has been detected. Some hunt for this dark matter, as they call it, to support their theory of the Big Crunch; while scientists like Penzias imply they might just as well hunt for unicorns.

A small number—including Arno Penzias, Arthur Schawlow, Charles Townes, and Allan Sandage—entertain the possibility of the existence of God. So, at times, did Paul Dirac, according to his widow. But even that leaves the whys unanswered.

What can they answer?

First the big picture:

Back in 1965 Allan Sandage, the dean of astronomers, had concluded the universe was closed and destined to move remorselessly toward the Big Crunch in some 80 billion years. That same year Penzias and Wilson discovered background cosmic radiation, impressive evidence for the Big Bang, and Sandage realized he'd mistaken white dwarf stars for quasars. He recalculated, but came to approximately the same closed-universe conclusion. Ten years later he changed his mind: the universe, he said, was open and would expand forever. And that's his present view.

Today [1995] in California, Sandage looks through a telescope at what he believes is the edge of the universe. That's also back to the time of the creation. "What is beyond that edge?" I asked Sandage. "Nothing," he replied. The universe hadn't yet begun to form.

But, of course, the universe is expanding into that nothingness. What's beyond that nothingess? Don't ask. Unless, of course, you can accept an infinity of nothingness as a plausible answer.

Sandage thinks of astronomy as "the impossible science," because it's based on opinions rather than the elusive facts: the objects of study being out of reach. So the astronomers' information rides "on beams of light from outer space," to be interpreted through ingenious instruments. The interpretation is where the disagreements start and the experts continue to argue about the age and size of the expanding universe.

No one agrees with the early cosmologists who thought it all began when God took the sun, moon, and stars out of a box and hung them up in the sky. Most now concede it started with a bang and is some billions of years old.

In either event, open or closed, every living thing is destined for extinction.

Unless of course, Robert Jastrow's race of immortals find a way to save the world. By all accounts—barring accidents—they'll have tens of billion of years to organize a celestial Noah's ark.

While cosmologists have come up with a plausible picture of the birth, growth, age, and size of the universe, particle physicists have been picking it to pieces and finding out how it works. By smashing particles together at immense speeds they produce neutrinos that, having no electrical charge and little if any mass, travel at almost the speed of light, penetrating, like Superman, everything in their path. Physics, as Wald points out, is full of such ghosts.

Murray Gell-Mann has named what might be the smallest particle "the quark"—with a nod to James Joyce. Almost everything is made out of quarks, smaller even than protons and neutrons, says Gell-Mann. And atom smashers in Europe and America proved he is probably right.

Today, physicist Victor Weisskopf feels that he and his colleagues have little left to discover about particle physics. If he were starting out again he would study a more tangible product of the Big Bang and the celestial furnaces than quarks and neutrinos: the human brain.

When I spoke with Wilder Penfield in the 1970s he implied there was little hope of consistently studying the brains of human beings while they were alive, that his observations had happened by chance during brain surgery on epileptic patients. Twenty years later Yale's Dr. Sally Shaywitz, a behavioral scientist, and her neurologist husband, Dr. Bennett Shaywitz, showed how it can be done without harming the subjects. With a team of researchers they used functional magnetic resonance imaging, a noninvasive method of scanning the active brains of 19 men and 19 women during which they noticed one way a woman can't be more like a man. Women use different parts of the brain to do similar and equally effective thinking. Women use both sides while men use just one side near the temple. When the announcement of this provocative discovery was made in February 1995, there were photographs to prove it.

I asked Dr. Bennett Shaywitz: What in fact is being shown?

Shaywitz: Differences in oxygenated versus deoxygenated blood in the brain. What you're looking at [in the photos] are differences in the use of oxygen in those areas. So, as people use that particular area of the brain to do a task they extract more oxygen from the blood. And that's what you image.

What was the task?

Shaywitz: They were thinking about the same task: to find and sound out the rhyming words in a list of nonsense words.

Has anyone suggested that the oxygenated blood and thinking may just be coincidental?

Shaywitz: Functional magnetic resonance imaging is a new technology, but there's an older technology that's been around for about 15 years. But it's the same principle. You're just measuring the metabolism of the brain in a somewhat different way. But it seems to be very well established that you are in fact looking at the way the brain is measuring. Obviously you have to have the tasks done very carefully. Because you are, in fact, correct. You want to make sure that the differences in metabolism reflect the tasks and not something just happening coincidentally.

Are there physiological differences between the male and female brain?

Shaywitz: That wasn't part of the study, but there are, I think, many studies showing women's brains to be smaller than mens, and brain size to be a reflection of body size.

[Autopsy reports show male brains are also more asymmetrical.]

You know it's said that we only use a small part of our brains. Might your studies indicate that women make more use of their brains than men do?

Shaywitz: I don't think you can extrapolate to that. Our interest is in figuring out reading and reading disability in children and young adults. And this was one of the first studies we've done using this technology. To try to figure out what areas of the brain are used when people read. And we were amazed to see this difference in men and women."[1]

Dr. Ruben C. Gur, director of the brain behavior laboratory at the University of Pennsylvania School of Medicine, backs up their study. He and colleagues used magnetic resonance imaging on the brains of 37 men and 24 women who were instructed to rest and not think of anything in particular. They found that men had higher brain activity in the ancient areas of the limbic system involved with action, and women in the newer and more complex parts of the limbic system associated with symbolic action. Dr. Gur explained the difference: action is when an angry dog bites; symbolic action is when it bares its fangs and growls.

Surprisingly, although it is our greatest resource, there never has been a concerted effort to study the brain by neuroscientists, psychologists, psychiatrists, and hypnotists—a CERN for brain research. CERN is the European organization for nuclear research in Geneva, Switzerland, which Steven Weinberg considers the greatest international scientific collaboration ever.

What astonished Einstein about the human brain was how it could understand so much about the world, and that mathematical equations were often the key to its secrets. But he never expected every secret to be discovered.

Scientists are still moving in the dark, trying to prove him wrong, occasionally colliding with one another, and what they take to be the light at the end of the tunnel proves to be at best a small cloud of fireflies.

But they know, too, that there's always hope. And that nothing is certain. In 1825 Auguste Comte said he was sure of at least one thing: the composition of stars would always remain a mystery. Some thirty years later Gustav Kirchhoff used spectrum analysis to show they were made of hydrogen and other gases. George Gamow, an early proponent of the Big-Bang theory was ridiculed for years and told his wild idea could never be tested. Then Penzias and Wilson came along with their discovery of background cosmic radiation and more recently the COBE satellite backed them up. In the summer of 1994 scientists discovered the top quark and the black hole—not by bringing them back alive but by indicating their existence.

Even Darwin's belief that survival of the fittest accounts for who and what's around today is under question. Harvard paleontologist Stephen Jay Gould suggests God does play dice and "debunks" according to Stephen Brewer, "the popular myth that the ascent of humans began with single cell organisms, moved through slimy fish and reptiles, from hunched hairy beasts to homo sapiens in progressive and orderly fashion. Instead, he says that evolution is a messy, risky, chancy business in which mass extinction can wipe out one highly successful species and spare another. 'You can be the most beautiful fish that ever swam,' says Gould. 'You can be perfectly equipped to survive. Then, one day the pond you live in dries up, and that's it, you die, no matter how fit you are.'"[2]

Nature seems to favor symmetry even in scientists: for every scientist to promote superstring there seems to be one to mock it. Michio Kaku, professor of theoretical physics at New York's City College, co-founder of string-field theory believes that brainpower alone can prove or disprove his theory and it in turn will be able to explain almost everything even the moment of creation, except of course, why? Still, that's another project worth investigating.

So that's where we are. Cosmologists can calculate to within a billionth of a second the universe's moment of birth and its probable death.

But no one is close to a scientific answer for any of the whys: why is there a universe? Why are we here?

Up and coming detectives need not despair. There's a lot left to discover. The mind is still largely a mystery. So that's one subject they can tackle.

As I write this on March 2, 1995, seven astronauts have just blasted off from Cape Canaveral. On this Endeavor mission they will use special telescopes to probe 10 billion light years away from Earth, in search of primordial hydrogen and an answer to the question "How did everything begin?" If they find hydrogen there, far out in space and back in time—what may be the initial byproduct of the birth of our universe—it will add to the accumulating evidence for the Big Bang and the expanding universe.

A few hours later, on this same day, while astronauts on the Endeavor were in orbit 220 miles above the Earth aligning their ultraviolet telescopes in space, two teams here on Earth at Fermi National Accelerator Laboratory announced they had independently found the elusive top quark. This fundamental building block of our universe is the most massive of all quarks [the others of varying smaller masses are named up, down, charm, strange, and bottom. Up has the smallest mass of 5 million electron-volts. Top has close to 200,000 million electron-volts.] Only the up and down quark is found in nature. The others apparently produced by the Big Bang disappeared from our visible universe, but have been recreated in particle accelerators. Dr. Boaz Klima, a leader of one Fermi team, said that the top quark is so astonishingly heavy that its decay may point to the existence of supersymmetric particles. If supersymmetric particles are discovered they might help achieve Einstein's dream of a unified theory by embracing gravity, the electromagnetic force, and the strong and weak nuclear forces, and answer the question: Are the properties of quarks determined by chance or by a fundamental unifying plan?

Mark Twain called humans the only animals that blushed or needed to. We're also the only animals to ask questions, despite the quizzical expressions on our dogs and cats.

No one could outdo Niels Bohr if we take him at his word that his every sentence—even what seemed to be a statement—was a question.

Einstein came close, saying: "The important thing is not to stop questioning. Curiosity has its own reason for existing. One cannot help but be in awe when he contemplates the mysteries of eternity, of life, of the marvelous structure of reality. It is enough if one tries merely to comprehend a little of this mystery every day. Never lose a holy curiosity."[3] He certainly followed his own advice.

The outcome of my putting questions to Nobel scientists and other luminaries is the urge to ask more, especially about the workings of the mind.

This is what Wald and Wheeler both speculate is our purpose on this whirling planet heading for infinity—to ask questions. Let's hope the final answers refute the doomsayers.

CHAPTER 1

1. Lidia Wasowicz, "Nobel Laureate Linus Pauling," *The Chicago Tribune,* April 18, 1985, Section 5, pp. 1, 3.

2. Anthony Serafini, *Linus Pauling: A Man and His Science* (New York: Paragon, 1989).

3. *Ibid,* pp. 14–15.

4. *Ibid,* p. 125

5. *Ibid,* pp. xvi, xvii.

6. *Life,* May 11, 1962, Vol. 15, No. 19.

7. Anne Sayre, *Rosalind Franklin and DNA* (New York: Norton, 1975).

8. Denis Brian, *Murderers and Other Friendly People: The Public and Private Worlds of Interviewers* (New York: McGraw-Hill, 1973) p. 209.

9. *Modern Medicine,* "Vitamin C in health and disease," July 1, 1976.

10. *Medical Tribune,* September 8, 1976, p. 11.

11. *Ibid,* p. 3.

12. *Ibid.*

13. George Wald and Linus Pauling *New York Times,* October 24, 1976, p. 14.

14. Serafini, p. 262.

15. Pauling letter to *Science News,* January 5, 1980.

16. Interview for WGBH-TV, Boston, June 1, 1977.

17. Florence Meiman White, *Linus Pauling: Scientist and Crusader* (New York: Walker, 1980).

18. *Linus Pauling: Crusading Scientist.* Produced and directed by Robert Richter, Corinth Films.

19. WGBH-TV, Boston.

20. *Bulletin of Atomic Scientists*, November 1979.

CHAPTER 2

1. James Gleick, *Genius: The Life and Science of Richard Feynman* (New York: Pantheon, 1992), p. 63.

2. Peter Wyden, *Day One: Before Hiroshima and After* (New York: Warner, 1985), p. 205.

3. Gleick, p. 222.

4. *Ibid.* pp. 292–293.

5. Victor Weisskopf, *The Joy of Insight: Passions of a Physicist* (New York: Basic, 1991), p. 134.

6. Richard P. Feynman, *OED: The Strange Theory of Light and Matter* (New Jersey: Princeton, 1985), p. 10.

7. *Los Angeles Times*, January 25, 1986.

8. Richard P. Feynman, *What Do You Care What Other People Think?* (New York: W. W. Norton, 1988), p. 185.

9. Richard P. Feynman, *Surely You're Joking, Mr. Feynman!* (New York: W. W. Norton, 1985), p. 342.

10. WGBH-TV, Boston, February 2, 1975.

11. Feynman, *What*, p. 244.

12. Gleick, *Genius*, p. 285.

13. Interview with Helen Tuck, August 11, 1993.

14. Wyden, *One*, p. 219.

15. Gleick, *Genius*, p. 434.

16. Feynman, *Surely*, p. 169.

17. Interview with Kip Thorne, August 11, 1993.

CHAPTER 3

1. Abraham Pais, *Niels Bohr's Times, In Physics, Philosophy and Polity* (New York: Clarendon, 1991), p. 296.

2. Ruth Moore, *Niels Bohr* (New York: Knopf, 1966).

3. Abraham Pais, *'Subtle is the Lord . . .' The Science and Life of Albert Einstein* (New York: Clarendon, 1982), p. 441.

4. *Ibid.*

5. Abraham Pais, *Inward Bound* (New York: Oxford, 1986).

6. Peter Goodchild, *J. Robert Oppenheimer, Shatterer of Worlds* (Boston: Houghton Mifflin, 1981), p. 21.

7. Banesh Hoffman with Helen Dukas, *Creator and Rebel: Albert Einstein* (New York: Viking, 1973), p. 193.

8. Werner Heisenberg, *Physics and Beyond, Encounters and Conversations* (New York: Harper and Row, 1971).

9. George Gamow, *Thirty Years That Shook Physics: The Story of Quantum Theory* (New York: Anchor, 1966), pp. 119–120.

10. *Ibid*, p. 122.

11. *Ibid.*

CHAPTER 4

1. Victor Weisskopf, *The Joy of Insight: Passions of a Physicist* (New York: Basic Books, 1980).

2. *New York Review of Books*, April 25, 1991.

3. *Insight*, pp. 133, 152–153.

4. *Ibid*, pp. 280–281.

5. *Ibid*, p. 131.

6. Thomas Powers, *Heisenberg's War: the Secret History of the German Bomb* (New York: Knopf/Cape, 1933); and David Cassidy, *Uncertainty: The Life and Science of Werner Heisenberg* (New York: Freeman, 1992).

7. *Nature*, May 27, 1993, p. 311.

8. *Ibid.*

9. Mark Walker, *Nazi Science: Myth, Truth, and the German Atomic Bomb* (New York: Plenum, 1995).

10. Gregg Easterbrook, "Are We Alone?" *The Atlantic Monthly*, (August 1988), p. 27.

11. James Gleick, *Genius: The Life and Science of Richard Feynman* (New York: Pantheon, 1992), p. 232.

12. *Insight*, pp. 87–88.

CHAPTER 5

1. Peter Wyden, *Day One: Before Hiroshima and After* (New York: Warner, 1985), pp. 50–51.

2. *Ibid*, pp. 114–115.

CHAPTER 6

1. Gary Zukav, *The Dancing Wu Li Masters: An Overview of the New Physics* (New York: Bantam, 1980).

2. Freeman Dyson, *Infinite in All Directions* (New York: HarperCollins, 1988).

CHAPTER 7

1. C. P. Snow, *The Two Cultures and a Second Look* (New York: Mentor, 1963).

2. Anthony Zee, *Fearful Symmetry: The Search for Beauty in Modern Physics* (New York: Macmillan, 1986).

CHAPTER 8

1. Meg Dooley, "Profiles: Arno Penzias, Managing Science," *Columbia*, December 1989, p. 37.

2. Jeremy Bernstein, *Three Degrees Above Zero: Bell Labs in the Information Age* (New York: Scribner's, 1984), p. 228.

3. *The Way of the Scientist* (New York: Simon & Schuster, 1966), p. 116.

4. *San Francisco Examiner*, July 2, 1978, p.7.

5. Malcolm W. Browne, "Despite New Data, Mysteries of Creation Persist," *New York Times*, May 12, 1992, p. C10.

6. Francis Crick, *Life Itself: Its Origin and Nature* (New York: Simon & Schuster, 1981), p. 158.

7. *Ibid.*

8. Meg Dooley, *Columbia*, December 1989, p. 37.

CHAPTER 9

1. *The Book of the Prophet Ezekiel*, Chapter One, verses 4–28.

CHAPTER 10

1. Theodore Berland, *The Scientific Life* (New York: Coward, 1962).

2. Richard F. Harris, *San Francisco Examiner*, March 4, 1985, p. A3.

CHAPTER 12

1. Jeremy Bernstein, "Profiles: Physicist," *New Yorker*, October 13, 1975, p. 96.

2. William L. Laurence, *The Hell Bomb* (New York: Knopf, 1951), pp. 6–7.

3. Harold C. Urey, *Current Biography*, 1941.

4. James Gleick, *Genius: The Life and Science of Richard Feynman* (New York: Pantheon, 1992), p. 144.

CHAPTER 16

1. Robert Stewart, reviewing *The Rise and Crisis of Psychoanalysis in the United States: Freud and the Americans, 1917–1986*, Nathan G. Hale Jr., (New York: Oxford University Press, 1995. *New York Times Book Review*, January 29, 1995), p. 13.

2. *A Certain Symmetry*, A. Carotenuto, Pantheon, 1984, p. 29.

3. "The Dora analysis was the first in which Freud, putting scientific interest above the principles of medical ethics, took it on himself to tell everything without even toning down the crudity of the words actually used." *The Psychoanalytical Revolution*, Marthe Robert (New York: Harcourt, Brace & World, 1966), p. 204.

CHAPTER 17

1. Forrest G. Robinson, *Love's Story Told: A Life of Henry A. Murray* (Boston: Harvard, 1993).

2. Claire Douglas, *Translate This Darkness, The Life of Christiana Morgan* (New York: Simon & Schuster, 1993).

3. Alfred Kazin, "Love at Harvard," *New York Review of Books*, January 28, 1993. pp. 3–5.

4. Anna Fels, "Hello, Jung Lovers," *New York Times Book Review*, February 21, 1993, p. 15.

CHAPTER 18

1. Morton Prince, *The Dissociation of a Personality* (Longman's, Green, 1906).

2. Flora Rheta Schreiber, *Sybil* (New York: Warner, 1973).

3. *Ibid.*

CHAPTER 19

1. Max Eastman, *Einstein, Trotsky, Hemingway, Freud and Other Great Companions* (New York: Collier Books, 1962), p. 24.

2. Ashley Montagu, *Immortality* (New York: Grove Press, 1955), pp. 66–67.

3. John E. Mack, M.D., *Abduction: Human Encounters with Aliens* (New York: Scribner's, 1994).

CHAPTER 20

1. Wilder Penfield, *The Mystery of the Mind* (New Jersey: Princeton, 1975), pp. xxviii, 4.

2. Richard Restak, M.D., *The Brain* (New York: Bantam, 1984).

3. Penfield, p. 115.

CHAPTER 21

1. Nicholas Wade, *New York Times*, April 20, 1994, p. D27.

2. Roger Sperry, "Structure and Significance of the Consciousness Revolution," *The Journal of Mind and Behavior* (Winter 1987), Volume 8, No. 1, pp. 37, 45.

3. *Ibid*, p. 55.

4. Nigel Calder, *The Mind of Man* (New York: Viking, 1970), p. 253.

UPDATE

1. Interview with Dorothy Munro, January 31, 1995.

2. *The Cambrian*, August 22, 1994.

3. David F. Salisbury, Stanford *Campus Report*, August–September, 1994.

4. *The Cambrian*.

5. *Denver Post*, August 23, 1994.

EPILOGUE

1. Interview with Dr. Bennett Shaywitz, February 21, 1995.

2. Stephen Brewer, *Columbia*, Winter 1994.

3. Personal memoir of William Miller, a *Life* magazine editor, May 2, 1955. *The Great Quotations* compiled by George Seldes, (New York: Pocket Books, 1967) p. 248.

Index